The Banff Centre for Management

ENVIRONMENTAL AUDITING

The Challenge of the 1990s

Sponsored in cooperation with The Banff Centre for Management by

Alberta Environment

Environment Canada

Alberta Environmental Research Trust

Gouvernement du Québec
Ministère de l'Environnement

Canadian Chemical Producers Association

Ontario Ministry of Environment

Canadian Petroleum Association

Petroleum Association for Conservation
of the Canadian Environment

The Banff Centre for Management

ENVIRONMENTAL AUDITING

The Challenge of the 1990s

edited by Felicity N. Edwards

University of Calgary Press

The University of Calgary Press
2500 University Drive NW
Calgary, Alberta
Canada T2N 1N4

Canadian Cataloguing in Publication Data

Main entry under title:
Environmental auditing

Proceedings of symposia held Feb. 27–28, 1990 in Calgary and May 29–30, 1990 in Ottawa.
ISBN 0-919813-92-5

1. Environmental auditing—Canada—Congresses.
I. Edwards, Felicity Nan. II. Banff Centre for Management.
TD194.7.E58 1992 658.4'08 C92-091565-5

Cover design by Jon Paine

Text designed by Terry Teskey

Printed and bound in Canada by Hignell Printing Limited.

∞ This book is printed on acid-free paper.

Contents

Appendices

The Banff Centre for Management and Environmental Auditing

THE BANFF CENTRE FOR MANAGEMENT is a residential, continuing education centre offering a wide variety of short-term, non-degree, professional development programs.

Our goal is to help managers function in an efficient, effective and harmonious manner in organizations which are being forced to adapt to a rapidly changing global environment. As such, our programs focus on developing creative leadership for managing change as well as initiating policy conferences on issues of contemporary importance.

It is with the above in mind that we developed, in conjunction with a variety of sponsors, Environmental Auditing: The Challenge of the 1990s, symposiums held in Calgary, Alberta, and Ottawa, Ontario.

Environmental auditing is now a priority for many resource-based industries, manufacturers, utilities and public sector organizations. The intent of the environmental auditing symposiums was to provide an initial forum for an international exchange of policies, practices and trends on environmental auditing. The following proceedings have been published with the intention of sharing this exchange with a wider audience. It is hoped by continuing to develop our awareness of the environment that we will meet this challenge squarely and with success.

Foreword

IN KEEPING WITH Environment Canada's policy of promoting environmental auditing as an internal management tool for both the public and private sectors, a federal/provincial/industry Steering Committee was formed in 1989 to organize two environmental auditing symposiums. It was felt these would provide an appropriate forum for the international exchange of environmental auditing policies, practices and trends. Sponsorship and guidance were provided by Environment Canada, the Provinces of Alberta, Ontario and Quebec, the Alberta Environmental Research Trust Fund, Canadian Chemical Producers' Association, Canadian Petroleum Products Institute and Canadian Petroleum Association.

The intent of these symposiums was twofold: to promote the practice of environmental auditing by industry and government in Canada, and to advance state-of-the-art of environmental auditing among practitioners. The consensus among participants was that both were fully realized.

The first symposium was held in Calgary, 27–28 February 1990, and the second in Ottawa on 29–30 May 1990. The material in this publication demonstrates the high quality of all presentations at the two sessions, and the workshop summaries included here will greatly assist in providing future direction for the practice of environmental auditing in Canada.

Environment Canada wishes to thank the members of the symposium Steering Committee for their time and effort in helping to organize the symposiums. Special thanks are extended to Mr. Peter Budzik of Peter T. Budzik and Associates as well as Ms. Felicity Edwards and Mr. Peter Hunt of The Banff Centre for Management for helping to shape the symposium program. Finally, we gratefully acknowledge the contributions of the speakers, session chairpersons and the participants towards the success of the symposiums.

A. Stelzig
Calgary Chair

J. Shrives
Ottawa Chair

Acknowledgements

THE DEVELOPMENT OF these successful symposiums would not have been possible without the hard work of the members of the Steering Committee. They participated in program development, as speakers and as resource people.

Appreciation is extended to the sponsoring organizations who helped to support both the Calgary and the Ottawa events.

Special thanks go to Art Stelzig and John Shrives who chaired the symposiums, to Elaine Smith for providing the speech summaries and to Linda Finn who typed the papers and produced the manuscript for the proceedings.

Felicity N. Edwards
Program Manager
The Banff Centre for Management

Symposium Steering Committee Members

Art Stelzig (Chairman)	Environment Canada
Michael Caranci	Ontario Ministry of the Environment
Guy Demers	Ministère de l'Environnement du Québec
Felicity Edwards	The Banff Centre for Management
Brian Hammond	Alberta Environment
Mark Lesky	Canadian Petroleum Association
John Reed	Petroleum Association for Conservation of the Canadian Environment
Don Robinson	Canadian Chemical Producers Association
John Shrives	Environment Canada
Ed Villeneuve	Noranda, Inc.

PERSPECTIVES

The Greening of Canada

G. Cornwall and B. Burns

IN FEWER THAN three centuries, humanity has managed to contaminate the entire global environment—the fouling of the nest, so to speak, has accelerated during the 20th century. This, coupled with virtually unrestricted resource consumption, global warming and the population explosion has brought us to a rather serious situation—potential extinction. Reputable environmentalists are now suggesting that the human race has all the characteristics of a cancer, the principal one being that, in the end, it destroys itself by killing the host. That may sound overly dramatic, but it does give us pause.

Environment Canada has been very active in recent years in promoting the voluntary use of audits by private concerns as well as government departments and agencies. This symposium and the one held last February in Calgary are the result of some of our efforts in this area. We are pleased to have been partners with industry and provincial agencies in this work. It is no exaggeration to state that there are two basic realities: (1) the environment is the basic commodity from which all else derives, and (2) Spaceship Earth is a closed, finite system—there is no escape!

In North America, today's environmental awareness was spawned in the sixties and carried through the seventies by the first Earth Day and the 1972 United Nations Environment Conference in Stockholm. It advanced in the eighties by a renewed concern for health and safety in general, as well as specific concerns related to toxic chemicals, hazardous wastes, recycling, waste management, environmentally safe products, etc.

G. Cornwall is Director of Federal Programs, Environmental Protection Directorate, Environment Canada, Ottawa. B. Burns is Regional Director General, Western and Northern Region, Environment Canada, Edmonton.

The "greening" of Canada has continued at a frantic pace through 1989 and has been reinforced by such things as landmark court decisions, heightened public pressure and environmental emergencies. For example, there were major oil spills, and there was major concern for ozone holes, St. Basile le Grand (PCBs), tainted fuels, drought, plight of fisheries, depletion of rainforests, and so on.

And if you thought 1989 was interesting let me predict that 1990 and beyond will be even more so! We have already had a second major tire fire, two major oil spills and unusually severe weather which has claimed several lives. We are also awaiting the resolution of such issues as the Oldman River Dam, James Bay II, etc.

We have reached the stage where an overwhelming majority of the public:

- is prepared to contribute to preserving the environment;
- wants to know more about the state of their environment;
- is concerned about environmental problems that affect health and safety;
- is prepared to change consumer habits;
- is becoming more active on environmental issues;
- is demanding action to protect the environment and expects development to proceed in an environmentally sustainable way;
- is concerned for a better quality of life for themselves and future generations;
- demands examination of alternatives; and
- wants less talk, more action.

Whether we represent industry, a government agency or an environmental group—whether the government agency be federal, provincial or municipal, *we must:*

- inform and educate the public;
- demonstrate environmental conscience and leadership through action;
- involve the public in solutions; and
- create new and better partnerships.

There is a belief that the 1990s will be a "litmus test"—a make it or break it time for society. We are at a crossroads and it is imperative that we begin to resolve some of the environmental issues which threaten our planet.

Environmental stewardship has arrived and is here to stay. We are turning to environmental auditing as a key element of environmental management. Regard it in the important context of pollution prevention—an essential ingredient of sustainable development.

Today and tomorrow, you will be examining this element of environmental management—one which can play a major part in meeting the challenge before us. It is not for me to go into the details of environmental auditing here; you will learn much more about all aspects of the subject from the excellent slate of speakers on the program.

I can tell you that environmental auditing has a special place in the enforcement and compliance policy which was issued when the Canadian Environmental Protection Act (CEPA) was passed in June 1988. The Enforcement and Compliance Policy was:

> To encourage the practice of environmental auditing, inspections and investigations under the Canadian Environmental Protection Act will be conducted in a manner which will not inhibit the practice or quality of auditing. Inspectors will not request environmental audit reports during routine inspections to verify compliance with the Act.
>
> Access to environmental audit reports may be required when inspectors or investigation specialists have reasonable grounds to believe that:
>
> - an offence has been committed;
> - the audit's findings will be relevant to the particular violation, necessary to its investigation and required as evidence;
> - the information being sought through the audit cannot be obtained from other sources through the exercise of the inspector's or investigation specialist's powers.

Environment Canada has been supportive of a number of environmental auditing initiatives undertaken by various industry associations and we commend them for their proactive approach. The Canadian Petroleum Association (CPA) has an excellent guidelines document for environmental auditing, for example. I might also point out that the historical perspective section of the CPA guideline has an interesting description of environmental auditing in the U.S. compared with the approach which is usually taken at Canadian facilities.

The federal government is committed to "putting the federal house in order." To that end, Environment Canada is developing a "code of environmental stewardship" for the Government of Canada. This code requires the government to achieve sustainable development both in its policy and program planning as well as in day-to-day operational decisions. Environmental auditing is seen as a key component of the Environmental Stewardship Policy. It is expected that departments and agencies of the Government of Canada will at appropriate intervals subject their operations to environmental audits and put in place milestone systems to ensure achievement of environmental goals by departmental managers.

We have encouraged and supported a number of environmental audit initiatives recently. An audit of "The Hill" was ordered by John Fraser and we are pleased to note that the Department of National Defence has audited a number of its bases. Transport Canada has begun an active audit program at some of its facilities and DOE has initiated a comprehensive environmental audit of departmental facilities.

In moving from a "react and cure" to an "anticipate and prevent" and "consumer to conserver" mindset, I see good environmental audit practices resulting in a number of benefits, including:

- reduced costs
- enhanced image
- improved commitment
- minimized liability
- prevention of pollution, and
- improved performance and compliance.

I'd like to explore a couple of these briefly.

Environmental Auditing Can Reduce Costs

- Prevention is far more cost efficient than remediation.
- Cost-benefit analyses, environmental impact assessments, spill prevention plans, decommissioning plans, etc. are tools which modern planners use to properly design new or dispose of old developments to avoid costly errors.
- Modern regulations are structured to prevent problems as opposed to simply setting out penalties.
- Environmental audits provide a means of evaluating how well company or governmental environmental goals and standards, developed during the planning elements noted above, are being implemented. They provide a basis for monitoring how well regulations are being followed and they can identify trends which require management attention.

- These outcomes translate into reduced costs through avoided court appearances and liabilities as well as through improved raw materials, product and waste handling practices.
- The costs of performing environmental audits will be compensated by financial benefits alone, not to speak of the environmental and human health and safety benefits.

Environmental Auditing Can Result in Enhanced Public Image

- As noted earlier, the public is prepared to change its consumer habits. Thus, there is likely some considerable benefit to be derived in terms of public image from the consistent and sound implementation of environmental audit practices. It is a good way to demonstrate good corporate citizenship.
- During the public review stage of any major project, companies and government departments make many statements about company standards and codes of practice. Environmental audits provide for an evaluation of how well these internal directives are being followed and a basis for initiating change where it is identified as necessary. This is extremely important. For example, an oil or pulp and paper company may have a multi-year development plan that could be seriously jeopardized by failure to honour publicly stated commitments.
- Those who fail to see the importance of this aspect, need only refer to recent media coverage of major pulp and paper developments. It is particularly true for large corporations with many product or service lines which can be negatively affected by a loss of trust in one small area, or as a result of activities far removed from the local market place.

Environmental Auditing Should Breed Improved Staff Commitment

- As often used as the statement may be, the fact is, managers are still "only as good as the people that work for them."
- The routine conduct of environmental audits, and even more importantly, any follow-up action identified, conveys the importance your agency attributes to environmental protection and preservation.

- This transfer of the corporate ethic has two positive results. The employees' thinking patterns are altered. Everyone, from the employee who thinks to stop continually running water during a cleanup operation to the middle manager who recognizes the risk in cutting corners on environmental problems, begins to factor the environment into his or her decisions and actions.
- The positive result is employee satisfaction and commitment. These days, with a large majority of individuals ranking the environment among the top two or three issues of concern, this is a significant benefit.
- Enhanced environmental awareness can do much to mitigate potential problems. Time and time again it is human error which underlies major environmental disasters.
- Clearly, the "green" employee is going to be proud to work for an organization that is demonstrating a sincere commitment to protecting and improving our environment. With this pride will come increased job satisfaction and the related productivity improvements.

Summary

- Environment Canada is committed to activity promoting environmental audits for the use by both private and public sectors.
- The Government of Canada is committed to ensuring that the environmental management of its departments and agencies is exemplary.
- Just a few of the significant benefits of performing environmental audits have been touched upon. Clearly there are others, as you will learn during this two-day session.

In conclusion, I would like to leave you with these thoughts: The complexity of the challenge before us will demand a contribution of every individual and require of us the best decisions that knowledge, commitment and technology allow. We must act now to ensure we are judged kindly by future generations.

Environmental Audits: A Corporate Necessity

J. Butler

THANK YOU FOR the invitation to participate in this conference. I'm pleased to have the opportunity to share with you my perspective on a critical process for the protection of our environment.

For Nova, this conference is timely. First, we have just completed the reorganization of the corporation and have fully integrated Polysar into the overall corporate structure. To mould this new corporation into a dynamic industrial leader which will achieve our vision of Nova, we must also be a leader in the management and stewardship of the environment. Failure to integrate environmental stewardship and economic development will derail our economic strategy and undermine our corporate objectives.

Second, we have recently strengthened Nova's health, safety and environment policy and implemented an environmental management system which commits the corporation to adhere to the regulatory requirements and go beyond compliance. A fundamental component of this system is the environmental audit program which has the full support of top management.

Nova, an Alberta Corporation

Before I try to grapple with our expectation of the environmental audit program, let me share with you a picture of Nova, a corporation which has undergone an exciting and challenging transformation in the past several years.

J. Butler is President of Nova Corporation of Alberta, Calgary.

Nova was incorporated in 1954 as the Alberta Gas Trunk Line Company Limited for the purpose of building, owning and operating a province-wide natural gas transportation system. Since the mid 1970s, Nova has expanded its businesses and has invested and sponsored ventures in several business areas, including: energy equipment manufacturing, cellular telecommunications, heavy truck manufacturing, pipeline systems for the transport of natural gas from the north, world-class polyethylene and ethylene facilities in Joffre, and the purchase of Polysar. We've also made a special effort to pursue investments which can be considered environmentally friendly. These include:

- Nova Plastics' Ecolyte, a plastic product which is degradable in sunlight;
- Hi Point Industries' Oclansorb, a biodegradable adsorbent thermally treated for the recovery of spilled products;
- Nova Geotech's polyethylene membranes for lining of landfills, ponds and lagoons; and
- greenhouses which rely on the waste heat generated by our facilities at Joffre and Princess.

In 1990, Noval Enterprises produced almost five million seedlings for reforestation and reclamation of mine sites and our search for new and effective environmentally friendly products continues. By far, however, Nova's greatest energy has and continues to be focused on additional resource development, expansion of the natural gas transportation system in Alberta and the development of a world-scale, internationally competitive petrochemical business. Nova, its subsidiaries and affiliates employ about 13,000 people in Canada, the United States, Europe and the Pacific Rim—countries with assets in excess of $8 billion.

The pipeline division, Alberta Gas Transmission, continues today to represent one of Nova's largest and most stable business operations. It is the primary transportation system through which natural gas is collected and delivered to Alberta border points for shipment to other markets. Nova transports more than 80 percent of the marketed Canadian gas production. Since the acquisition of Polysar in 1988, the petrochemical-related businesses have been restructured into three divisions:

1. Nova Petrochemicals: which produces a broad range of products: olefins, aromatics, fuel products and methanol from plant sites in Joffre, Medicine Hat, Corunna and Sarnia.

2. Nova Plastics: with plants in Joffre, Cambridge, Montreal, Springfield, Leominster, Marysville, Moore, Copley, Decatur and West Haven—produces linear low density polyethylene, low density and high density polyethylene, polystyrene and polypropylene and performance-oriented products such as acrylic resins and geomembranes.
3. Nova Rubber: which produces synthetic rubber at plants in Strasbourg, Antwerp, Orange and Sarnia.

In addition to the above divisions, Nova owns 43 percent of Husky Oil Ltd. and a variety of smaller companies involved in exploration, production and manufacturing. And at any point in time, Nova is storing and transporting feedstock and products throughout the world.

Need for Audits

The sensitivity to and awareness of environmental problems is having a profound influence upon the way we and other companies are operating our businesses. While the 1980s saw the heightened concern of the public, industry and the government towards environmental and health related issues, the 1990s will see the greening of industrial facilities and operations throughout the developed world. The CCPA's Responsible Care Program and the CPA's Environmental Code of Practice are two leading examples of our industry's dedication and commitment to manage our activities without upsetting the delicate balance of our environment. Nova thoroughly endorses these association initiatives.

We acknowledge our responsibility to act in the interest of the environment and we acknowledge that our every action will be subjected to increased scrutiny during the 1990s. California's Proposition 65 heralds the forthcoming regulatory philosophy and the challenge. It is designed to eliminate detectable discharges into the environment. If elimination is not feasible, industry will be required to demonstrate that there will be no significant risk to the environment or the health and safety of the public.

An impact may have local, regional or global implications. The impact may have been carefully studied and understood, or it may be of such serious potential consequence that corrective action may be initiated without a proven causal link such as with global warming. Regardless, the generator will be required to modify processes, reduce resource utilization, reduce, recycle and reclaim wastes and upon occasion eliminate certain products from the market place. The action may be in response to regulatory pressures or, more likely, a desire to eliminate unacceptable risks.

This is not a prescription for bankruptcy. However, to integrate environmental stewardship and financial objectives will represent a formidable challenge to Nova and the industry. To pursue the challenge effectively, we must pursue environmental management in a systematic and thorough fashion.

We must have knowledge of the contaminants we are introducing into the environment and understand their implications. We must record and communicate our concerns to the people who must act to remediate the environmental hazards identified in any of our plants, facilities or business activities. We must follow-up and ensure the corrective action is properly completed and is effective. And most important, we must anticipate problems and take corrective action before environmental problems occur. As E.S. Woolard, Chairman of du Pont, accurately pointed out in his speech to the American Chamber of Commerce in May: "avoiding environmental incidents remains the single greatest imperative facing industry today."

The Audit Process

The environmental audit identifies and assesses past, present and potential environmental problems, verifies compliance with regulations and corporate standards and assures corrective action is implemented. As such, good environmental management and performance begins with the conduct of thorough, consistent and independent audits based upon accepted standards.

Each company has a different set of objectives for the pursuit of its environmental audit program. For Nova the objectives are multifaceted. However, we've set our two key objectives which are derived from our desire to improve our environmental performance and regulatory compliance and to ensure the environment is protected and the health and safety of our employees, the community and the public is maintained. I've previously alluded to both:

1. *Compliance:* Our program must verify and document compliance with government regulations, corporate policies and standards, and good operating practices. Discrepancies from requirements and deficiencies must be identified and corrective action programs must be developed.

2. *Risk Assessment:* Our program must identify and evaluate existing and potential environmental liabilities or significant hazards.

Our strategic plan demands that we anticipate and avoid costly surprises related to environmental incidents which may interrupt operations and/or threaten the health and safety of our employees and the public. There are numerous other objectives which are secondary to these and they have been elaborated upon by many individuals. Many are natural spinoffs of an environmental audit program. The most obvious is the assurance to management that operations are consistently satisfying regulatory requirements, corporate policy and good operating procedures. Without an audit process it is difficult to envision a system which would demonstrate that legal and ethical responsibilities are properly being addressed and that due diligence is being performed.

The audit program by its very nature reinforces the corporate standards. During major acquisitions, such as the Polysar takeover, the corporate environmental audit program becomes a means of communicating expectations and training personnel in the corporate philosophy and vision. In the long-run, the environmental audit program will identify current and anticipated costs, enabling resources to be effectively applied and environmental control programs implemented in a cost-effective manner.

Maximizing the Benefits of the Audit Process

I would like to conclude my discussion on environmental audits with a note of caution. Audit programs are conducted to identify and assess situations which may be contrary to regulatory requirements, corporate standards or represent unacceptable risks to the corporation, community or public. They are constructive management tools without any intent to reprimand or reward individuals in the corporation.

Both the U.S. EPA and Canadian government strongly encourage industry to adopt environmental auditing as a management tool for similar reasons—to help achieve and maintain compliance with environmental laws and regulations, to identify and correct environmental hazards and to improve environmental performance. In the process of identification, however, there is a risk that our own information could be self-incriminated and provide information in support of legal proceedings.

Even though both governments have indicated audits will not be used for witch hunts or accessed during routine inspections, there is no assurance that audits will be held confidential and inaccessible under the authority of a search warrant or through the legal discovery process.

At present, few if any audits have been used in pursuit of prosecution. However, if government agencies and the public access audits for the purpose of prosecution, companies will seek effective means of protecting the information. Access to the information will be restricted, thereby limiting the successful application of a critical environmental management tool.

Conclusion

Unequivocally, a driving force behind the conduct of audits is the desire of the corporation to abide by regulatory requirements and to identify risks to the environment. In a situation such as we find ourselves at Nova, the auditing process is a critical tool. It enables us to establish priorities and promptly implement corrective action to ensure that each of our divisions and subsidiaries is in compliance with all relevant regulations and corporate policy.

I leave you today with my full support for your efforts to improve the environmental audit process. And I challenge you to look beyond simple compliance and ensure that the environmental audit process continues to emphasize communication in the broadest sense and involves every individual who can assist in the improvement of environmental performance.

Environmental Auditing: The Ontario Government Perspective

R.A. Findlay

I AM PLEASED to bring the greetings of Ontario Environment Minister Jim Bradley, who sends his regrets that he cannot be here with you today. I hope you will accept me as his humble alternative—it is indeed a pleasure for me to be here—to have this opportunity to say hello to some good friends and colleagues with whom I have enjoyed working in the past and who are here today. But I am *particularly* pleased to have this opportunity to speak with you about two issues which are particularly timely, appropriate and *connected.* They are sustainable development and environmental auditing. I want to explore these connections. I want to make the case that environmental auditing is a key concept in moving Ontario forward to this somewhat difficult-to-define goal called sustainability. It would seem logical that I start by describing sustainable development, and then I will look at environment auditing, and in the process, look at the connections.

What is Sustainable Development?

Sustainable development is an issue, a collection of ideas, actually, that interests me very much—and excites me about the prospects for our future. Our *common* future, to borrow a well-used phrase, and the future of your children and mine. I'd like to briefly look at the concept of sustainable development in terms of where it came from, what it could mean, why it's possible, and just maybe what we can do to get it—in terms of auditing. I believe this is worth doing because some people

R.A. Findlay is Director of the Ontario Round Table on the Environment and the Economy.

suspect that sustainable development may be the latest fad. And if sustainable development is a fad, then I submit that environmental auditing is a hula hoop! Everybody's doing it and we're going around in circles! Both terms—sustainable development and environmental auditing—are being tossed around by people who perhaps aren't too clear about the ideas themselves, or who are likely to interpret such words in a way that reflects their own narrowly defined self interests. I'll do my best to persuade those of you who are among the sceptics and perhaps I won't succeed. But I won't be too concerned if some of you remain sceptical. First, because I value critical attitudes. Second, because being a fad for a while might not be a bad thing. It may be the way an idea gets noticed and accepted, a sort of rite of passage on the way into daily use. A point I hope we are rapidly approaching.

The concept of developing sustainability is anything but new. In fact many economic historians have commented that it was the traditional form of development prior to the industrial revolution some 200 years ago. Well, I'm not an economic historian. I prefer to believe that the concept of developing sustainability is just good common sense. I believe that concepts of sustainability have been suppressed in the human psyche for a long time. And perhaps it is these beliefs that lead to my optimism about the future. My optimism that we will be able to unlock this suppressed but latent instinct for survival that we hold. And that we will indeed have a sustainable future. I believe that the answers to the question "What is sustainable development?" are all about us. Rather like the fellow in Mark Twain who was amazed to find he had been speaking prose all his life and didn't know it.

Natives, conservationists and some environmentalists have been basing their thinking and actions on "sustainable development-like" ideas for a long time now. And I think that it's interesting to note that ideas about making the connections between our environment and our economy were becoming clear in 1977 *Conserver Society Report* of the Science Council of Canada and very prominent in the 1984 McDonald Royal Commission on the Economic Union and Development Prospects for Canada. But the phrase really began to receive special attention when it became a focus, maybe *the* focus, of the views expressed in the 1987 *Report of the World Commission on Environment and Development.*

The Brundtland Report gives a very simple definition of sustainable development: "meeting the needs of the present without compromising the ability of future generations to meet their own needs." This is a goal with which few would quarrel. But then, by itself, it doesn't help much either, because, for the Brundtland Commission, the concept had a much larger significance. Within their broad and comprehensive view of the world, its members saw as never before a wide range of interlocked and disastrous global trends: in population, food security, ecosystem degradation, energy consumption, industrial pollution, urban overcrowding, the greenhouse effect, the ozone layer, and more. When considered together, these trends showed the Commission that the world is on a course that cannot be sustained—certainly not indefinitely, and in some cases not even for much longer.

The Commission members believed that the wasteful exploitation and consumption in wealthy countries and the poverty and scarcity in poorer ones were no longer sustainable. Indeed, both pose real and significant threats to global security. When viewed in a "big picture" way, these practices are now seen to be caused by short-sighted, and often well-meaning but needlessly damaging and wasteful forms of human activity. In short, unsustainable development. It is interesting that the Commission did not adopt an anti-development stance. Whatever may have been the relationship between development and environment in the past, the Commission decided from now on that environmental conservation and protection is and would continue to be critically *dependent* on continued economic development. But not the kind of economic development we have experienced in the past, the *right* kind, much *smarter* development, the kind that our environment can sustain. In fact, sustainability has two sides: economic and environmental. Sustainable development occupies the common ground that human beings can afford and the environment can accept. The concept implies that such common ground exists and must be pursued.

So where does all this leave us? Are we better off with or without the term? I'd like to add one word to the concept and carry on from there. If we consider those activities and alternatives which are *more* sustainable, I think we can individually and collectively make significant progress by using this concept. As part of a transitional strategy, the key for me is to think in relative rather than absolute terms, but to have up there, as a target, as a kind of summit at the top of the mountain, the goal of sustainable society.

The Brundtland Commission express confidence that human society is not only capable of moving to more sustainable kinds of development but that it is absolutely inevitable—if the planet is to survive. It expresses confidence that we *are* in a position to reverse the devastating global trends. The Commission made many recommendations to help get the solutions under way. Running through the recommendations is a central theme: the need to bring environmental considerations from the edges to the centre of economic decision making, to the centre of our planning systems and to the decisions we make about the use of this particularly blessed part of the world.

Ontario Round Table

Many of you may know something about the Ontario Round Table. Chaired by Murray Elston, Chairman of the Management Board of Cabinet, it reports to the Premier, and is composed of six cabinet ministers, six CEOs, a mayor, a labour leader, a native leader, an academic and members of environmental organizations. The priority activity of the Round Table is to oversee the production of a provincial strategy or plan to achieve sustainable development. The process we will employ will be open and consultative, and will serve as an educational tool for those both directly and indirectly involved with the strategy's preparation. At present, we are finalizing the first of two Challenge Papers which will serve to seed the public debate on what our strategy should aim to achieve and over what time frame.

Our first Challenge Paper will constitute an *indicative agenda* for Ontario. It will identify a number of preliminary challenges in six environmental issue areas that are directed at six macro-sectors of our economy. We will seek feedback on the agenda to indicate where you agree or disagree, where people think it should be tightened, and where it should be modified or added to.

The Round Table has identified six principles which help define sustainable development and which provide the foundation for our strategy development process. I want to focus on a couple of these principles for a minute, by way of introducing my remarks about environmental auditing. The first principle is *anticipate and prevent.* Anticipate and prevent is better than react and cure. In other words, pollution is expensive. Prevention is much cheaper.

Environmental auditing fits—because it is a proactive approach, not a reactive one.

A second principle of sustainable development which the Round Table has identified is called *informed decision making.* Auditing establishes environmental awareness at every level of an operation, from process operator to CEO, and recognizes the responsibility which industry has toward the society in which it operates. It makes it clear that this responsibility goes beyond that required by our traditional so-called "back end" regulations, which are often a react-and-cure response to a problem. Maybe, in some cases, a perceived problem—dare I say?

How Does Auditing Fit in with Ontario's Environmental Agenda?

Ontario views environmental auditing as an important step in our transition to a sustainable society. Environmental auditing places responsibility for environmental protection on those whose activities may impair our natural surroundings. It builds in environmental protection as a *cost* of operating. Incidentally, the third Round Table principle of sustainable development is full-cost accounting. Again, note the fit. Actually, all jurisdictions these days are moving in the direction of greater environmental protection. Studies conducted for the Ontario Ministry of the Environment conclude that any competitive advantages gained from lax controls may be short lived.

Auditing is consonant with a *shift* in the Ministry of the Environment's "polluter pays" principle to a new "prevention pays" principle. Speaking of shifts, we see a trend in shifting inspection and assessment activities from ministry staff to industry. Ontario is already moving in this direction. The best example is the province's Municipal-Industrial Strategy for Abatement (MISA) which requires nine industrial sectors to monitor their toxic discharges to Ontario waterways in preparation for abatement regulations. The ministry will continue to conduct random audits as part of this approach.

Environmental auditing allows businesses to know what they need to remedy to meet Ontario's standards and guidelines. If you do a thorough audit, you will be well positioned to meet tougher limits set in the future. A comprehensive audit is the best way to stay abreast or ahead of aggressive environmental protection programs. Not to put too fine a point on it, environmental auditing is a fine way to keep your boss out of jail!

Looking to the Future

Ontario recognizes the need for a clearly defined, proactive approach. The Round Table is developing a sustainable development strategy. The Ministry of the Environment is developing a discussion paper on auditing, soliciting input from all interested parties. Ontario is dealing with unresolved issues in the following three areas: guidelines, voluntary versus mandatory auditing and the confidentiality of auditing documents.

Let's look at:

1. *Guidelines*

- In the absence of clear guidance to industry, environmental audits will be inconsistent and incomplete.
- A comprehensive audit requires detailed assessment of technical, operational and legal aspects.

A guideline should be based on the need for industry at the corporate level to take the broadest possible view of the connection between operations and their effects on the environment. In this context, audits should assess:

- company environmental policy and how it is communicated to staff;
- employee roles, responsibilities and training as each affects compliance;
- status of compliance with acts and regulations; and
- potential for non-compliance as a result of equipment failure.

2. *The Voluntary versus Mandatory Question*

- There are two ways of approaching this issue. One encourages voluntary participation in the auditing process by pointing out its benefits to industry. This could include a firm due diligence defence.
- The other approach contends that a voluntary system already *de facto* exists and that audits should be required as a reactive complement to abatement programs.
- The resolution of this stumbling block may lie in developing a regulated auditing requirement contingent on some trigger mechanism.

3. *The Third Point is Confidentiality*

- To date, this has been the most difficult issue to resolve. I understand that *industry* is firmly opposed to carrying out environmental audits as long as the environment ministry retains the right to seize the audit document as evidence of non-compliance.

- While the Ministry understands industry's reluctance here, Ontario is equally uncomfortable with giving blanket confidentiality and immunity to these documents.

Calm, constructive, cooperative approaches are obviously needed to find the common ground. Let's not forget about the principle of *informed decision making.* While several key items must be ironed out before a regulated audit process is established, Ontario is committed to working with all interested parties to find solutions.

In closing, I believe that sustainable development is *the* context—environmental auditing is a tool. It is a good tool. An educational tool. A team-building tool. To a craftsman, a good tool is a good investment, not a cost. So is environmental auditing. It is good, *proactive* management and an essential part of sound environmental management. Somebody once said the problem with the future is that it "ain't what it used to be." Let me give you my own personal glimpse. Soon, we will be expanding the scope of environmental auditing to include products and processes and procurement patterns as we seek discipline on the path to sustainable development. The bottom line is this: sustainable development is the context. Environmental auditing is the tool.

Environmental Auditing: The Corporate Perspective

W.N. Kissick

I WOULD LIKE to begin by quoting Ed Woolard, chairman of the world-wide Du Pont organization, who observed in a recent speech that those of us in industry have to develop a stronger awareness of ourselves as environmentalists. These days, he said, we cannot concede the name "environmentalist" to any single group of spokespeople. In that context, while I am here in my role as a CEO, I believe those initials stand for Chief Environmental Officer as well as Chief Executive Officer. That feeling is shared by my colleagues who are guiding Canada's companies into this new decade.

Our knowledge about our environment and its protection is steadily growing and we are in better shape than ever to turn our commitment into decisive actions—actions that will move us toward a closer harmony between continued economic growth on the one hand, and excellent environmental stewardship and sustained quality of life on the other. That, I think, is a goal we all share.

Now I recognize that my mandate is to discuss environmental auditing; however, I thought you might also be interested in taking a look at the broader picture of environmental stewardship in the industrial chemical industry—a stewardship program that provides a solid foundation for performance auditing. It's no secret that the chemical industry is viewed in a very poor light by a large section of the Canadian public.

Much of the present mistrust and fear stems from misconceptions of what motivates us and how we behave toward the environment. In the past, we viewed ourselves as "the invisible industry." At the time, this was regarded as appropriate. The industry was diligently working behind

W.N. Kissick is Chairman and CEO of Union Carbide Canada Ltd., Toronto.

the scenes, with little hoopla, to provide chemicals that helped to feed, shelter and clothe the growing human race, and improve the health and life expectancy of the global population.

Intensified concerns about the future of our environment made that posture obsolete. In a new climate of environmental awareness, Canadians were demanding to know how chemicals affected them; they wished to be told about the risks involved in chemical manufacture. And they wanted to decide for themselves whether the benefits outweighed those risks. Quite suddenly, our low profile became a detriment. People tend to fear what they don't understand and they did not comprehend the chemical industry.

Today, we are doing our level best to prove to Canadians that we have our house in order through a broad-ranging program called responsible care. The program is being administered by the Canadian Chemical Producer's Association (CCPA), an organization made up of 74 chemical companies with more than 200 plants in about 115 communities coast to coast. These plants produce chemicals worth $11 billion a year, which represents more than 90 percent of all the chemicals made in Canada.

As a Director of CCPA and Chairman of its Responsible Care Management Committee, I am deeply involved in making this program a way of life for chemical enterprises. And let me assure you the program has teeth. For example, as a condition of membership in CCPA, each company must commit to maintaining the responsible care standards and to review and improve its safety practices and procedures in all aspects of its business. This means ensuring the safe and responsible management of chemicals through all stages of their life cycle—from their inception in the laboratory through to their ultimate disposal. It also means keeping the community surrounding every plant fully informed about what goes on in the plant, and what would be done if something went wrong.

With considerable input from the public, CCPA has developed six codes of practice. These codes cover the areas of research and development, manufacturing, transportation, product stewardship, waste management, and community awareness and emergency response. The codes lay down very specific performance requirements in each area. They are not optional. Each member company must have written policies, procedures, programs and auditing for each element. And each company is required to develop an implementation schedule for complying with all the codes. Its progress will be measured against that schedule by CCPA, which will not only keep tabs on the progress being made, it will report industry performance publicly. In this way, Canadians will be able to monitor the industry and judge us against our commitments.

You might be interested to know that a community advisory panel, which provided valuable input in developing these codes, included representatives of environmental interest groups. We believed that the understanding and acceptance of our toughest critics was essential if the codes were to stand up to public scrutiny.

Responsible care represents the framework on which we are building a sustained, long-term effort to achieve the highest standard of environmental performance, and to maintain an open, candid dialogue with Canadians. We are accountable to all of our stakeholders. We want to demonstrate that, and to restore public trust and confidence in our industry. And, I should add, that the responsible care effort of the Canadian industry is setting a standard for the world. Similar programs have already been adopted by the chemical industries in the United States, Britain, Australia and New Zealand. And responsible care is influencing the industries of other countries, too.

CCPA is actively promoting the message of responsible care throughout the world. A commitment to this type of program by chemical associations around the globe would demonstrate that the chemical industry is founded on ethical principles and intends to keep on living up to them, wherever in the world it may operate. This, then, is how the chemical industry has progressed in its drive for excellence in health, safety and environmental protection. Obviously, an essential part of that thrust is the continuous monitoring of our performance, and identification of potential problem areas, through in-depth auditing.

The industry itself, through CCPA, has developed a very detailed approach to safety assessment. This takes the form of an extensive questionnaire and rating system that allows companies to rank every aspect of their safety performance from the role of management to the design of equipment, from maintenance and security to emergency planning and community communications; and from responsible use of the industry's products to proper disposition of the industry's wastes. Comprehensive as it is, this is strictly self-assessment. Defining and correcting specific deficiencies is the responsibility of each company. Given the climate of public opinion, and the tough standards which the industry has imposed on itself, state-of-the-art auditing of performance is of paramount importance.

Let me explain a bit about Union Carbide's auditing effort, which is undertaken in a uniform way throughout Carbide's world-wide organization. The Board of Directors of our parent Union Carbide Corporation, has appointed a Health, Safety and Environmental Affairs Committee which, by the way, is chaired by a director who is also the chairman of the World Wildlife Foundation and the Conservation Foundation. In Union Carbide Canada, our entire board serves as the HS & EA committee. These board committees review and approve all policies and standards with the objective of an HS & EA performance that is "second to none." The auditing process is one of the committees' major responsibilities.

Audit teams, consisting of three or more staff specialists and often an outside consultant as well, regularly visit and inspect all operating and administrative sites. The teams review all functions for compliance with corporate and legislated standards, including the management process, occupational and general health, personnel safety, environmental protection, product stewardship, waste disposal, and community awareness and emergency response. These audits are generally conducted with minimal advance notice to local management.

The auditors operate completely independently of other functions and businesses. This way, they can go about their work with a totally unbiased, objective view. They report to a corporate Vice-President who has direct access to the Chairman and the Board Committee. The fundamental job of each of our audit teams is to make sure we are complying with our own internal safety requirements, and all the appropriate government laws and regulations. And, if the team ever determines that a problem needs immediate attention, it has the power to shut down the operation on the spot.

Teams audit all locations at regular intervals. Initially, the frequency of these audits depended on the degree of hazard that was revealed by a comprehensive screening process. This ranking included such factors as: operational hazard; population density both within the plant and in its immediate proximity; local topography; distance to local emergency services; volume of potentially hazardous materials contained on site; and potential for such natural disasters as hurricanes or earthquakes. Based on worst-case scenarios, plants were categorized around the world according to their inherent and potential risk of deficiency. This risk assessment determined the initial audit priority and frequency.

The audit teams prepare comprehensive reports following each audit. The reports note all deficiencies from corporate or governmental standards, and also note observations on management or operational practices which might inhibit future compliance. A draft report is reviewed with local management to ensure accuracy but not to debate the substance of the audit findings.

When a final report is issued, the facility is rated against a "full-compliance" benchmark. Local management is required to provide within two months specific action plans to achieve correction of deficiencies and respond to the observations. These action plans are tracked to completion by senior management and the audit group. Satisfactory resolution of deficiencies is confirmed in subsequent audits.

The audit report and rating are the premises for several management actions:

- determination of resources and schedule required to correct deficiencies
- assessment of the performance of the facility management
- adjustment of the future audit frequency.

Although the process I have described is the cornerstone of our auditing program, it is supported by a number of other approaches which address specific issues with equal rigor:

- A number of formal audits by engineering and operating personnel take place in the course of building new facilities. Documented reviews are carried out at the completion of process design, just prior to the commissioning of a new facility and again when the facility reaches stable operation at design rates.
- Operational audits are conducted regularly to be certain that critical facilities are being operated in accordance with design criteria and best operating practices. These audits also test prior risk assessments to determine if any key parameters have changed.
- And, finally, we also ask our operating people to perform continuous but less formal audits of their plants on a week-to-week basis, particularly to look for mechanical or environmental changes which might lead to increased risk.

In all cases, any deficiencies are noted and formal action plans are implemented to correct them.

I am totally committed to the auditing process. In my opinion, it is an essential ingredient of an effective program of environmental stewardship. But it should not be restricted to environmental protection alone. I believe that intensive auditing should always be directed at occupational health and personnel safety, as well as protection of the environment. Indeed, a thorough but realistic auditing process has helped to maintain Carbide's health, safety and environmental procedures at a level that is, according to the Boston consulting firm Arthur D. Little, more thorough and comprehensive than those of any of our competitors.

Finally, I'd like to extract a few principles of the Carbide auditing program that, I suggest, might make a good foundation for anyone contemplating their own program.

1. The standards against which the audit is conducted must be objective and technically valid—they cannot be arbitrary. Aiming for an unattainable ideal is a frustrating exercise. Setting sights on achievable and justified goals is what keeps people involved and produces results.
2. Standards must have received the understanding and acceptance of the line organization. Getting the line organization on board and supportive should be one of the very first steps.
3. The audit team must include expertise in all of the relevant disciplines. You can't audit something you don't fully understand. The integrity and professionalism of the team members must be apparent to all.
4. The auditing team must be independent of line management and must have unrestricted access to the senior executive level when necessary. In other words, it must be able to operate as an independent body with authority to cause action appropriate to their findings.
5. The audits must be fair and reasonable and should avoid any frivolous nit-picking.
6. The auditing process itself should not be static. As we expand our technical knowledge, as we develop new and better methods of risk assessment and, as our society strives to define its cost/risk criteria, we must be prepared to refine and adapt.

There is still an enormous amount to learn, which is why a symposium like this is so vital to all of us.

Environmental Auditing: A Management Process

P.T. Budzik

IN MY PRESENTATION, I have the opportunity to set the stage for what will be taking place during this symposium/workshop. I would like to examine some fundamental concepts surrounding environmental auditing and explore the answers to three basic questions, namely:

1. What is an environmental audit?
2. How does it fit into the general framework of management practice?
3. What are the reasons why today, more than ever before, private sector companies and public sector agencies should embark on environmental auditing programs?

What is an Environmental Audit?

The practice of environmental auditing came into being in the early 1970s in the United States, principally in reaction to tough U.S. environmental legislation exacting stiff penalties against corporate officers for their corporation's failure to meet statutory requirements. Based largely on the positive side benefits of the U.S. experience with environmental auditing, corporations with holdings in Canada introduced the concept of environmental auditing to their Canadian subsidiaries in the early 1980s.

The need for environmental audits really grew out of the increased complexity and scope of environmental legislation aimed at controlling emissions to air, water and soil from stationary, mobile, non-point and

P.T. Budzik is President of Peter Budzik & Associates Inc., Bethany, Ontario.

point sources, and the control of chemical and physical agents from cradle to grave and beyond.

In an attempt to better manage their environmental performance in the face of this complexity, companies borrowed a page from a well-established business control tool known as the financial audit. By applying a similar process to that used in auditing of financial transactions to environmental practices, senior company executives could improve their compliance assurance level.

The word *audit* has its roots in the latin word "auditus," or hearing, which, to the Romans, meant an official examination and verification of accounts and records. An environmental audit has been defined by the International Chamber of Commerce as:

> A management tool comprising a systematic, documented, periodic and objective evaluation of how well the organization's management and equipment is performing in meeting the goal of safeguarding the environment by: (*i*) facilitating management control of environmental practices; and (*ii*) assessing compliance with company policies, which would include meeting regulatory requirements.

In a more simple and less rigorous form it can be viewed as "a voluntary, systematic and objective *evaluation* and *verification* of a company's compliance with external environmental legal requirements, and its own internal environmental policies and standards." I would like to bring the following characteristics of this definition to your attention:

1. It is company initiated. It is not something imposed on an organization from an outside agency, i.e., it is voluntary.
2. It calls for a systematic approach with pre-planning and forethought, designed to meet specific goals and objectives.
3. It calls for an evaluation of the findings by application of measurement techniques and technical judgement.
4. It requires verification of the findings.

By implication this definition also calls for:

1. An honest and open working relationship between the auditors and auditees in order to arrive at a true assessment without fear of punishment.
2. A system of documentation and record keeping of the auditors' working papers and final reports.

3. A mechanism for identifying deficiencies and striving for continuous improvement in environmental performance.

I am often asked how an environmental audit differs from an environmental impact assessment, or, for that matter, an environmental compliance monitoring program applied to industrial operations.

An environmental impact assessment is a study carried out during the early planning stages of a new project. It generally takes into account the existing economic, social and physical environment at a proposed site location. An estimate is made of the changes that will result to these environmental sectors from the installation and operation of the new facility, such as a gas plant or oil refinery. It also identifies the design requirements necessary to meet compliance with accepted regulatory standards.

By contrast, an environmental audit applies to existing organizations and operations and is designed to give management an independent look at its organizational and systems strengths and weaknesses.

On the comparison with environmental compliance monitoring the situation is somewhat different. Both deal with existing operating facilities.

Most facilities, as part of their internal control system, sample and test their emissions and waste materials to ensure that they are in compliance with their environmental permit limits and other statutory requirements. If an exceedance is noted, corrective action is initiated to bring the operation into line with the legal limits.

Environmental auditing, on the other hand, looks beyond the day-to-day reporting and checking of test results. It observes and analyses the organizational structures and the management control systems in place to make sure that the correct parameters are being monitored and that the proper equipment, operating and maintenance procedures are in place to keep the facility well within specifications set by both government and company management.

Environmental Auditing as Part of the Management System

This brings me to the second objective of my presentation, i.e., to demonstrate that environmental auditing is really simply a logical progression evolving from sound environmental management.

The key word here is *management*. In classical terms *management* is generally viewed as a series of functions such as planning, organizing, staffing, directing and controlling the operations of an organization.

Since external factors affecting these management functions such as market demands, raw material pricing, applicable legislation, are constantly changing, the system itself requires continual review and realignment to cope with the new circumstances.

The process therefore can be viewed as a cycle of activities involving the setting of goals and standards, developing plans to achieve the goals, implementing the planned actions, measuring and evaluating the effects of the actions, correcting any existing deviations from the goals and identifying new opportunities for improved future performance. The cycle returns to the point of origin, and, based on new data inputs, establishes revised goals and objectives and continues on from there.

Environmental auditing is intended to bring in "fresh eyes" to examine all the facets of this cycle. In the strict sense, it is an expansion of management's control function, only this time it is applied to the overall cycle. As such, it measures the strengths and weaknesses of the environmental management system in place and points the way to upgrading its environmental performance.

Just like the carrying out of an environmental impact assessment study is considered good engineering practice for major projects, so too is instituting an environmental audit program considered to be good management practice.

Although the principal reasons given for carrying out environmental audits are the reduced risk of tort liability and ensuring compliance with applicable legal requirements, there are other reasons as well. These include:

1. Identifying poor operating and maintenance practices at a plant.
2. Assessing the effectiveness of existing environmental control plant equipment and identifying opportunities for cost reduction and reduction of pollution.
3. Avoiding "surprises," resulting in unexpected and costly litigation or clean-up requirements.
4. Improving overall compliance assurance.
5. Improving employee morale through raising environmental awareness.
6. Enhancing short- and long-term environmental planning.

Continuous Improvement

Today, because of increased international economic competition, many managements, who are keeping their pulse-finger on productivity, are also looking at instilling a company philosophy of continuous improvement. This philosophy espouses constant striving for improving product quality, customer satisfaction, profitability, productivity, resource use and environmental and occupational health and safety performance.

If it is to be successful, a concept like continuous improvement must be amenable to goal setting and measurement. As a refinery manager once told me, "If you don't measure what you've done, you will never be able to plan what to do." Environmental auditing helps make this possible.

The culture of constant improvement does not come naturally into a workplace. It has to be promoted from the highest levels within the organization. But a good environmental audit program can assist in fostering a climate of continuous improvement.

Why an Environmental Audit Program is Important Today

For the past ten years pollsters have been examining the social climate of this country. Politicians and corporations have been watching the results. The outcome of the most recent polls indicate that 1990s is the decade of the environment. In an article published in the 1 January 1990 issue of *Maclean's* magazine, the environment was named as the most important issue facing the country, topping the GST and the national deficit.

The article went on to say "the [public] concern about the environment was in stark contrast to the past two years when free trade dominated Canadians' view of the national agenda, while the environment was cited by only 2 percent and 10 percent of respondents in 1987 and 1988 respectively." Private industry and government have been perceived to be doing a poor job of protecting the environment by over half the Canadian population. In response, governments have been getting tougher with environmental offenders. In Ontario prosecutions for environmental offenses have more than quadrupled and convictions have more than tripled over the past five years. This trend is predicted to continue along with jail sentences for senior executives and company personnel found guilty of gross environmental negligence.

As Jon Grant, president of Quaker Oats Co. of Canada and a strong proponent of environmental audits, recently said, "You're going to have to do it [environmental audit] as if you were doing a financial audit. Public expectations are too high to get away with rhetoric instead of action."

In addition, as many companies who have embarked on audit programs have come to realize, apart from creating public good will, environmental auditing is simply an instrument of good business practice in the face of escalating clean-up costs and environmental remediation. I would strongly advise those of you who have not yet embarked on an environmental auditing program to give it serious consideration.

Summary

In closing, I would like to say that we have examined environmental auditing and showed how it differs from an environmental impact assessment and an in-house environmental compliance program. It is a voluntary program of evaluation and verification of compliance with internal company policies, procedures and standards and government regulations. As defined, an environmental audit is really an extension of management's control function applied to the entire management cycle. Both because of increased public expectations and more stringent regulatory requirements, companies and agencies are well advised to adopt environmental auditing programs to enhance their environmental performance.

I would like to quote Robert D. Kennedy, the Chairman of Union Carbide Corporation, who, in addressing senior chemical industry executives, said:

> There can be no status quo when it comes to our environmental performance. There can and there should be a state of creative dissatisfaction that leads to improvement in emission reduction, in waste generation, in safe transport and handling of chemicals in public communication and ultimately in public perception.

Environmental Auditing: Canadian Industry Perspectives

J.W. Reed

FEW ISSUES TODAY hold greater consequence for corporate credibility than those related to the environment. As part of their overall approach to environmental protection, many organizations in the public and private sector now apply a management practice known as environmental auditing.

Environmental auditing is an important topic because it has the potential to make a significant contribution to environmental protection. It is also an exciting topic because the field is growing quickly as more and more organizations adopt programs.

Many meanings are ascribed to the term environmental auditing. In many respects, the field is more confusing today than it was five years ago. This confusion is a reflection, in part, of the differing uses and applications of auditing, giving rise to such distinctions as:

- compliance vs. management audits
- self vs. third party audits
- top down vs. bottom up audits
- comprehensive vs. selective audits, etc.

It is also a reflection of the growing number of parties with a perceived stake in the development and practice of auditing, including private industry, governments, public interest groups, labour, academia and consultants. Each of these stakeholders views environmental auditing in fundamentally different ways.

J.W. Reed is Coordinator, Environment and Safety, McColl Frontenac Inc., Don Mills, Ontario.

The objective of this paper is to provide an overview of environmental auditing as a tool for environmental and risk management and to provide a brief summary of the Canadian context, in terms of both current activities as well as future developments. This understanding is important since environmental auditing is sure to become the subject of increasing industry and government activity.

Description of Environmental Auditing

Environmental auditing is a management tool used to monitor performance and detect where risks and problems might exist. As a process for self-evaluation of environmental performance, it is a deliberate attempt by an organization to periodically and objectively examine operating locations, assess the effectiveness of local practices, and verify conformance with pre-defined performance expectations, including regulatory compliance. Environmental auditing programs:

- are internally driven
- are voluntary
- have a formal structure and organizational status
- use vigorous and systematic procedures
- are conducted on a regular cycle, and
- are close-looped.

Environmental auditing is like other types of internal auditing: it is concerned not just with end-point performance, but with the spectrum of activities that create the end-point performance. For example, internal financial auditing isn't concerned only with balance sheets and cash flow—it looks at the entire set of practices that create and maintain sound financial management. Similarly, internal safety auditing isn't concerned only with accident frequency rates and loss time injury rates—it looks at the whole set of systems that create and maintain safe working conditions. And so it is with environmental auditing: same idea, different topic.

It is important to note that environmental auditing is a *check* on existing management systems and controls. It does not replace such systems and controls. An audit is simply a diagnostic tool. It only has value when there is something to diagnose. Many people fall into the trap of believing that an audit is a panacea, a cure-all, a program that will somehow magically improve environmental management. That is patently false: an audit is just the check; ultimately, it is the implementation of proper systems and controls that affects environmental performance.

Three other points are salient. First, the golden rule for environmental auditors is that they must be independent from the activities that are being audited. If they are not, objectively and credibility are jeopardized. Often, this means the audit is conducted by individuals external to the audited facility, such as Head Office personnel, Division personnel, personnel from other operating locations within the company, or even external consultants. Second, a commitment to respond to the findings of the site audit in a reasonable fashion and in a reasonable timetable is an intrinsic part of any auditing program. In addition to due diligence requirements, follow-up assists in developing and installing the basic management systems and controls in a facility operation. Third, an individual facility audit cannot exist in a vacuum: it is part of a broader environmental auditing program. An environmental auditing program also includes, for example, assigned responsibilities, defined performance expectations, pre-determined audit frequency and site prioritization, auditor training, reporting procedures, action planning and (as discussed above) follow up mechanisms. Moreover, an auditing program is but one component of environmental management and is only one of many tools that are available to today's environmental professionals. It is a good tool for some companies, but it isn't right for every company. Indeed, it is not needed for every company: environmental management systems can still be effective without an auditing program.

Evolution of Environmental Auditing

Environmental auditing was not suddenly "invented" per se. In effect, the early pioneers simply recognized that the "tried and true" practice of internal auditing could have benefit in the environmental area. Some U.S. companies started environmental auditing programs in the early 1970s. While the practice experienced steady growth in the private sector in the ensuing years, the popularity of environmental auditing exploded in the late 1970s and the early 1980s, as soon as governments, consultants and lawyers entered the arena. From that point on, the spread of auditing was rather predictable, as conferences were held, journal articles published, various forums for discussion established, training programs developed, and so forth.

Two distinct trends had an influence on the early U.S. environmental auditing pioneers:

1. One was increasing concern about the extensive liabilities and risks created in the environmental area. In addition to the large amount of statute law and regulations, there were also civil and tort liabilities, costs of potential incidents, fines and penalties, and corporate image impacts to consider. A tool was needed to provide some protection against those liabilities, i.e., to identify and manage risks before they became a problem.
2. The other was a shift in environmental management from a "macro" to a "micro" focus. In other words, after making large capital investments for basic pollution control hardware, companies needed a tool that would help fine-tune the operation and assist in the development of management systems for environmental excellence.

The influence of these two trends was not mutually exclusive; however, a good portion of U.S. programs were developed in response to the trend towards increased liabilities, and were thus designed mainly as protective mechanisms. Such programs were necessary given the overabundance of regulations and the confrontational nature of regulators.

In Canada, the situation was somewhat different. A 1984 survey of Canadian industry conducted by the author for the federal Environmental Protection Service revealed that 25 percent of survey respondents had auditing programs in place. Most (but not all) of those programs were designed primarily as a tool for management excellence, with a much smaller emphasis on protection against regulatory non-compliance. The relative lack of environmental legislation allowed companies to focus on improving the management and control of plant activities that had an impact on the environment. In effect, some companies were using audits as a driver to develop and improve internal standards. At that time, environmental auditing was a "quiet" practice, with no government involvement. Today, given the explosion of regulations in Canada, the more adversarial relationship between industry and government and the emphasis on enforcement and penalization, and given industry's apprehension over the potential access to voluntary audits by government, the situation is quite different: More and more programs are being designed as or modified to be protection oriented.

Program Characteristics

At a macro level, environmental auditing programs share a number of similarities and have certain common characteristics. These include:

objectives, scope, techniques, program organization and structure, reporting and follow-up, protection of audit information, etc.

At a micro level, however, there is a substantial amount of variation among corporate environmental auditing programs, as such programs are generally tailor-fit to suit the specific needs, resources and culture of the individual firm. Just as each firm is unique, so too is each auditing program.

This paper will discuss objectives, scope, techniques and reporting in order to illustrate the range of programs that exist.

Objectives

The 1984 survey illustrated that companies institute environmental auditing programs for a variety of reasons. In fact, some twenty different program objectives were identified. These ranged from:

- providing management assistance to operating entities
- eliminating surprises
- verifying compliance with regulations
- avoiding fines and damage claims
- enhancing employee awareness
- improving cost control.

Although most auditing programs have multiple objectives serving a number of clients, there is normally one *over-riding* driving force and one *key* client. This driving force is the major determinant of the overall shape of the program. It will affect:

- how it is staffed and resources (what level of independence is needed, who the auditors are, etc.)
- what techniques are used
- which facilities are audited
- who the findings are reported to, etc.

In some auditing programs, the principle driving force is to verify for company management that the environmental affairs of the company are being managed appropriately *and* that the firm is in compliance with regulations. This is the "protective" mechanism described earlier. The principle client is often the senior management in the corporation. Some people describe these as "police audits," implying an "over the shoulder," "big brother" approach.

At the other end of the spectrum are environmental auditing programs that are primarily intended as a line management tool. In this mode, the audit is a service to help line managers improve their management system. Here, the approach is more cooperative, and less confrontational. We find a heavy emphasis on assessing the effectiveness and adequacy of environmental protection activities, as opposed to verifying compliance with standards. In point of fact, most programs tend to blend some aspects of both of these approaches, rarely pursuing one to the exclusion of the other.

Scope

Auditing programs vary tremendously in scope. Scope can be thought of in many ways. One way is by considering *what* you look at. The "what" can be broken down into *subject matter* and *function*. Typical audit subject matters include air management, water management, waste management, emergency response and spill control, hazardous materials, process units, health, external relations, and others.

Some auditing programs are "comprehensive" in their treatment of subject matter. In other words, they address a full slate of subject matters such as those listed above. Other programs are more selective in their approach, looking at only one or two subjects from the full slate. The selective approach allows for a more in-depth analysis of a given subject matter, all other things being equal, whereas the comprehensive approach gives a much wider perspective on plant performance. Both approaches are valid, and both are used by Canadian firms.

Another way of looking at scope is by function. For example, within each specific subject matter there are a number of parameters to be examined. These include permits and regulatory requirements, company policies, operating practices and procedures, management systems, monitoring and record-keeping practices, training programs, employee awareness, equipment maintenance, and others.

Some auditing programs choose to focus only on regulatory requirements for each subject matter reviewed. (Some people call these "compliance audits.") Other programs put an emphasis on the management systems and controls that lead to compliance with regulations. (Some people call these "management audits.") Other programs still attempt to look at all aspects of environmental management, including the full list of functions. Many programs start out as so-called compliance audits, however, as the programs mature, the focus often shifts towards the management systems which underlie compliance.

Another way of looking at scope is in terms of *where* you audit. Some programs look at only those activities occurring within the plant fenceline. Other programs extend to division and head office staff departments. Others still audit outside the company, including, for example, off-site waste disposers, haulers, contractors, etc.

Techniques

Audit programs vary in the techniques they employ. Essentially, environmental auditing is a process for collecting and verifying information. The 1984 survey revealed that auditors structure their information gathering in a number of ways. For example, they:

- use formal protocols, checklists and questionnaires
- watch people as they work
- conduct interviews with employees and management
- physically inspect units and facilities
- take photographs
- examine records and files, and, in some cases
- send samples to external labs.

Reporting and Follow-Up

Clear documentation and effective communication of audit results is a key step in the audit process. Most audit programs involve preparation of the written report following the site audit, although the specific procedures (format, content, roles and timing) are program specific. Variation occurs in terms of:

- who prepares the written report
- who reviews draft copies
- who receives final copies
- who prepares corrective recommendations and whether they appear in the written report
- whether deficiencies only are noted in the report, or both deficiencies and strengths

- what the role of legal counsel is
- whether legal protections are invoked
- how quickly reports are prepared and processed.

Once identified in the reporting process, a deficiency must be addressed. For this reason, follow-up procedures are a cornerstone feature of most audit programs. Follow-up involves developing and implementing a plan for correcting priority findings on a defined schedule (as well as, in some cases, providing a rationale for not taken action against certain findings), monitoring the implementation of the plan and validating the effectiveness and adequacy of implemented corrective actions. Follow-up procedures vary considerably in terms of assigned responsibilities, degree of formality, documentation, scheduling, legal protections, etc. Concerns over due diligence require that follow-up procedures be well defined and understood.

There are other characteristics to describe auditing programs. But the point is that there is a tremendous variety to this program, labelled environmental auditing. The characteristics of programs are unique to each company having been designed to meet the specific management needs, resources, and culture of the implementing firm.

Benefits of Auditing

The 1984 survey of Canadian environmental auditing practices indicated that a full 25 percent of the respondents were conducting auditing programs. The number of firms conducting environmental auditing programs has increased since then. The fact is, Canadian firms are voluntarily adopting environmental auditing programs. Obviously, there must be some benefits to the process. These benefits generally derive from both *the product of auditing* (i.e., the results) as well as from *the process* used to arrive at the product.

The chief benefits include:

1. Measurement of achievement and progress in environmental protection. As an on-going monitoring tool, audits provide a benchmark against which environmental programs can be measured.
2. Substantiated data on compliance with external requirements, company policies and standards and good management practices. A solid data basis critical to effective decision-making; audits provide a database that can be used by many management levels.

3. Protection against liabilities and proof of due diligence. Audits help to identify weaknesses in the systems that underlie compliance. Audits also provide documentation of due diligence.
4. Identification of risk exposures and early warning of potential problems. As the saying goes, "An ounce of prevention is worth a pound of cure."
5. Improved communications between and within operating facilities. Audits lead to the sharing of information throughout an organization.
6. Education and training. The sheer visibility and process of audits speak volumes to line personnel.

Clearly, audits are instituted for sound business reasons. But there is an added plus. Audits can result in a better level of protection for the environment.

Activities/Development

The field of environmental auditing is a dynamic one and has undergone dramatic changes in the past five years. The following section is intended to sample some of the recent activities and expected developments in the field, with a particular emphasis on the Canadian experience.

Industry

Since the 1984 study, individual companies as well as a number of business associations have been active, to varying degrees, on a number of fronts. To begin with, the implementation of environmental auditing has experienced steady and continuing growth within the Canadian private sector, cutting across a range of industry sectors and company sizes. Many companies have a formal general environmental policy which includes a commitment to conduct environmental audits; some companies have even developed specific policies on environmental auditing.

In addition, many industry associations have developed, or are in the process of developing, environmental policies to govern and support the affairs of members. Several industry associations have specifically endorsed the use of environmental auditing and are actively promoting an awareness of auditing through internal projects, newsletters, sponsorship of conferences, speaking engagements, etc. The Canadian Petroleum Association (CPA), through its "Environmental Code of Practice," specifically commits its members to conduct environmental audits. Similarly, the Mining Association of Canada's (MAC) "Environmental Policy" endorses self-monitoring techniques.

The development of guidelines for environmental auditing is another area of intense activity. Most companies have prepared formal procedures or manuals which describe the "how to" and "what to" of their auditing programs. Many industry associations have also been active in this area. Examples include the CPA's "Environmental Audit Guidelines for the Petroleum Industry," the Canadian Chemical Producers Association (CCPA) "Self-Assessment Program," and the International Loss Control Institute's (ILCI) "Environmental Protection Module." Internationally, the International Chamber of Commerce has developed a position paper on environmental auditing.

Despite the proliferation of environmental regulations, there are many industry activities/practices that affect environmental performance but for which written, agreed upon, standards do not exist. Consequently a large number of companies are active in developing internal environmental "standards," which in many cases go beyond the requirements of law. Several industry associations have also embarked on related programs. Examples include the Petroleum Association for the Conservation of the Canadian Environment's (PACE) "Environmental Performance Measures," CPA's "Environmental Operating Guidelines," and CCPA's "Responsible Care Program."

Many companies have developed their own internal standards of auditing, e.g., use of formal protocols, auditor qualifications and training, report format, etc. There is some reluctance to formalize auditing standards at an industry association level for fear of restricting the flexibility companies desire to adapt to unique circumstances. Attempts by other stakeholders to standardize audits are discussed later in the paper.

Aside from these key activity areas, several trends or developments within industry are noteworthy:

- Due to ever-increasing corporate and individual liabilities, senior managers increasingly see themselves as the key clients in auditing programs. This has resulted in a shift to "top down," protection-oriented programs where the emphasis is on both demonstrating compliance as well as assuring continued compliance.
- Greater linkages between the auditing function with risk management/insurance functions.
- Several companies already conduct environmental audits of selected locations beyond the fenceline. This is likely to continue with the

extension of audits to include outside services, contract haulers, waste disposers and suppliers. In some cases, audits are included part of contractual obligations.

- Due to fears of "toxic real estate," pre-purchase (physical) assessments will extend to broader pre-purchase environmental audits.

Government

The actions of government in Canada and internationally, have had and will continue to have a dramatic direct and indirect effect on the development and use of environmental auditing. Indirectly, a business environment characterized by an increase in the number of extensiveness of environmental regulations, strict enforcement and prosecution, and stiff penalties and fines, heightens the need for auditing-like tools. As discussed earlier, such a business climate creates a defensive orientation.

In a direct sense, federal, provincial and municipal government departments and agencies are examining auditing as both a potential public policy tool (i.e., as an instrument to promote public benefits) as well as a potential management tool within the public sector. In this area, governments are grappling with a twofold question: Can private sector auditing lead to public benefits? and, if so, what role should government play to enhance such benefits?

Several different policy options have been and are being examined by various jurisdictions in Canada. These options include:

- promoting the concept
- encouraging private application of auditing and development of state of the art tools
- mandating the use of audits (by statute or regulation)
- gaining access to audit information for the purposes of enforcing environmental regulations and prosecuting violations
- incorporating audits into routine abatement activities, such as the development of control orders, etc.
- conducting "audits" of companies.

Some of these options have the potential to debilitate the development of auditing on a voluntary basis, and for this reason, have generated heated debate by and within the private sector. The challenge for government is to maximize public benefits while not inhibiting the development and use of auditing on a voluntary basis. Environment Canada, as part of

CEPA, released a formal policy on environmental auditing that goes a long way to meeting this challenge. Several provincial governments are known to be preparing subject policies. Moreover, environmental auditing is being discussed in a number of quasi-government forums, including advisory councils, roundtables, etc.

Several levels of government in Canada are preparing to implement auditing programs internally. Efforts by Environment Canada, Canada Post, Transport Canada and the Department of National Defense are notable in this regard. It is expected that other government departments, crown corporations and municipalities will follow in some measure.

General

Beyond those activities specific to government and industry, perhaps the most striking general observation to be made is the emergence and growing profile of new "stakeholders" in the environmental auditing field. Each of professional consultants, academic institutions, lawyers, organized labour, politicians and public interest groups are now active, in their own way, in the field. Each has a different perceived stake as reflected in their needs and values and each is exerting an influence on the development of the practice. For some of these groups, auditing is viewed as a tool to fight polluters, for others it is a chance to promote sound environmental management, for others still it represents a business opportunity.

This is a healthy trend in the sense that auditing is being promoted and discussed in many circles. The vigorous discussion and debate is sure to offer new perspectives which may enhance everyone's understanding of the concept and may at the same time lead to improvements and creative applications. Moreover, the promotion of auditing through conferences, journal articles, training courses, etc. is unprecedented. On the other hand, we must be on guard to ensure that the self-interest of stakeholders and the pressure they exert does not become a detriment or disservice to the practice of environmental auditing.

Another general trend, related somewhat to the above, concerns the simmering debate regarding environmental auditing as a public policy tool versus a private management tool. This debate is sure to heat up in the near future. While proponents on both sides have legitimate points of view, finding the appropriate balance, if at all possible, is a difficult challenge. Certain of the public policy options under discussion will negatively affect the development and application of auditing in private organizations, where the voluntary practice is already producing laudable results.

A final noteworthy trend is the move toward institutionalization of auditing. In the broadest sense, this refers to standardizing auditing and auditors. Several formal forums on environmental auditing already exist. In the U.S. the Environmental Auditing Roundtable and the Institute for Environmental Auditors are examples. A Canadian Environmental Auditing Roundtable has also been formed.

There have also been discussions around the accreditation of auditors—indeed in some U.S. states such a process already exists. Other activities already described, such as the development of guidelines and principles of auditing, are further examples of institutionalizing auditing.

Conclusion

Since its earliest application, the goals and functions of audit programs have expanded considerably. It is no longer restricted to measuring compliance and monitoring performance. It is used effectively as a tool for line management, education training and communication. As such, it has tremendous potential to make valuable contributions to protection of the environment.

Bibliography

Reed, J.W. (1984). *Environmental Auditing in the Canadian Private Sector.* Ottawa: Environmental Protection Service.

Pre-auditing Procedures in the Department of National Defence

T. Downs and K. Kavanagh

FIRST OF ALL I want to emphasize that the Department of National Defence (DND) has taken a general management approach to environmental audits, or as we call them in DND, environmental baseline studies. With the continual battle for person years in the government we have decided to rely on the private sector for expertise in designing and carrying out audits and preparing reports with recommendations for corrective action. It is then up to DND management to ensure that an action plan is prepared to carry out the recommendations and to monitor implementation of the action plan. Environmental audits of Canadian Forces bases were started in 1984. The aim was to have an expert do a complete inspection of a base and advise on the day-to-day impact of that base's activity on the environment and how we comply with current regulations and guidelines. The intent of the inspection was to advise on any shortcomings so they could be rectified. In this way the base or establishment would have a reasonable baseline on which to plan its subsequent activities, which presumably would be carried out in an environmentally conscious manner. This would ensure that the base was operating, and would continue to operate, in an environmentally safe and legal manner.

The baseline studies were so named for obvious reasons. It was also thought that a commander would welcome such a study whereas he might not exhibit the same cooperation toward an audit of his functions. Semantics aside, the aim is to assist. No effort is made at all to criticize or penalize bases for non-conforming situations. They are directed to correct adverse situations and get on with their jobs. They do so with an

T. Downs is Director, Conservation and Environment, National Defence Headquarters, Ottawa. K. Kavanagh is Head, Environment and Safety Engineering, Conservation and Environment, Department of National Defence, Ottawa.

increased sense of environmental awareness, but without having to worry about past sins.

Policy and Procedures

It is National Defence policy to complete audits on all Canadian Forces bases and installations. The most significant problem is the time available to get the studies done since we can only manage two or three a year from National Defence headquarters. There are thirty bases in Canada and many smaller installations. To date, eight bases have been completed with three out to contract now. This program will be greatly accelerated to finish the initial screening in three years.

The audits are conducted using consulting firms. The brief-in documents for each consultant contract are continually updated by the Directorate, Conservation and Environment, based on the results of the most recent studies as well as the latest regulatory guidelines. In this way the quality of the studies is continually improved.

Costs for these studies have been increasing since we started, partly because of the progressive additions to the scope of work. The latest study for a small base domestic site was $35,000; for a normal base with a large training area $92,000; and for a large industrial base in the order of $200,000. Protocol development only for special and complex facilities has cost about $75K.

A copy of each completed study is provided to the local district or regional office of Environment Canada partially in recognition of support provided during the study. Input from these agencies has been beneficial in improving the reports and in implementing any mitigation measures.

Objectives

The early studies were done on bases with small or no ranges or training areas. Canadian Forces bases are complete miniature municipalities with a wide variety of functions and activities that must be examined, from industrial workshops to sewage treatment plants. Recently, the studies have been expanded to involve major ranges and training areas. Here, the physical impact of military training on terrain, wildlife and vegetation is examined. The base's military mission will greatly affect the content of the studies, be it an army, airforce or navy base. However, the study must look at anything with a potential impact on the environment.

However, although we have concentrated on the environmental impact of our activities we are naturally moving more towards compliance audits. Non-regulated activities will continue to be assessed as to our compliance with accepted codes of good practice or as to their ethical or moral obligations—in other words, in order to maintain good corporate citizenship. The latter objective, while it may seem like "motherhood," is or should be a very real aim of any organization. It sets the stage for increased public confidence, not only in on-going activities, but in new proposals that may have to be assessed in future.

In summary then, the objectives of the DND audit program are:

1. To ensure compliance with federal legislation/regulation and provincial legislation where appropriate (it is DND's policy to comply with the strictest standards, regardless of legal jurisdiction);
2. To ensure compliance with internal policies and regulations;
3. To ensure all non-regulated activities are carried out in an environmentally sound manner, and to assess the risks associated with those activities (that is, to ensure they are in accordance with accepted codes of good practice);
4. To ensure that all activities are carried out in an environmentally ethical manner (that is, as a good corporate citizen);
5. To ensure all activities are carried out in accordance with the best principles of waste reduction, recycling of products, reuse of otherwise discardable products and recovery of resources for reuse or resale;
6. To ensure that the environmental management system is in place and followed to comply with regulations and good practice on a day-to-day basis; and
7. To reassure senior management and the public that we are operating in a safe and legal manner.

The reports must make recommendations for improvement of:

1. activities/operations;
2. departmental policies and regulations as they apply to the establishment being audited; and
3. management systems as they apply to the establishment being audited.

The audit reports must be very user friendly, for effective management use, as well as for field personnel who will be implementing the recommendations. In some cases, further detailed study is undertaken where a potentially significant problem is identified, but the determination of extent and degree of significance is outside the scope of the baseline study. For instance, discovery of fuel oil contamination in soil surrounding a fuel tank farm may be identified by the study but the type and extent of the contamination would be the subject of a follow-on study to permit a proper assessment and subsequent clean up of the area.

Protocol Development

Long before the site visit there are a number of steps to be taken. Protocol development is, of course, the key to any successful audit. This development can be done especially for a facility, if it is a special-purpose facility or one of extreme complexity. Otherwise, a generic protocol, suited to application at many similar facilities can be used. In DND, a generic protocol has been used for baseline studies. This has been successful, as long as it has been progressively improved with each study and obvious additions for special formations were made. Recent audits scheduled for defense research establishments were obvious candidates for special protocol development. A consultant, Acres International, was contracted to produce these protocols and develop the specific audit procedures.

Pre-audit Procedures

Part of protocol development is to determine what subjects are to be included in the environmental audit, that is, the scope of the audit. Based on trying to attain the audit objectives mentioned earlier, a very comprehensive list of subjects would seem necessary. While the following list applies generally to Canadian Forces bases and units, the list must be tailored to the establishment being audited, as it may be only one part of a larger organization. Generally, the following subjects are included (these subject headings would form the basis for the protocol outline):

1. Environmental management system
   - policies
   - compliance with Environmental Assessment Review Process, CEPA, WHMIS
   - environmental training

2. Air pollution control
3. Solid waste disposal
4. Water supply and water pollution control
5. Hazardous materials management (cradle to grave approach, including hazardous waste disposal)
6. Emergency response to spills
7. Occupational health and safety (WHMIS)
8. Natural and cultural resources management
9. Noise pollution
10. Fuel storage and handling
11. Bio-medical waste disposal

A key factor, particularly in DND audits, is to try and minimize the amount of workload put on the establishment to be audited. Filling out of a pre-visit questionnaire and the appointment of an on-site coordinator are now considered essential. It is up to the team, with some assistance from the establishment, to dig up the information they need. Any other approach will likely result in less than satisfactory cooperation from the unit in question, resulting in a less valuable audit.

A pre-audit visit, with the headquarters people involved, is necessary to explain the purpose and methodology of the audit and to explain the questionnaire. This one-day visit will pay dividends in accuracy of information received as well as cooperation of the individuals concerned.

As to who should compose the audit team, many factors should be considered. First, is the expertise available in-house? In DND, it is not available for this purpose. Therefore a technical manager must oversee the audit team and the experts are hired (i.e., a consulting firm). Again, experts in the activities of the unit are not necessary. The expertise should be engineering, science, environmental technology, generally. In fact, communication skills and a keen sense of observation are the prime requisites. The responsibilities of each member of the team and especially the team leader must, of course, be well documented before commencing the on-site audit.

In summary, the pre-audit steps are:

1. organize the audit team;
2. review the establishment organization and activities;
3. scope the audit;
4. develop the detailed protocols or adapt generic protocols;
5. finalize the audit team composition;
6. prepare the pre-audit visit questionnaires;

7. make the pre-audit visit;
8. review applicable policies, regulations and codes of practice;
9. review any previous audits;
10. finalize the scope and audit protocols; and
11. develop the detailed agenda for the audit visit.

Note: The order of these steps can be varied to suit the facility and audit team leader's methodology.

Concluding Remarks

Both the audit protocol and the auditors must be flexible in their approach. Unforeseen problems will inevitably crop up that require the protocol to be modified. Also, the auditee must never feel threatened. The aims or objectives of the audit as outlined earlier must be explained as many times as necessary. The overall aim being to correct or improve the situation so all personnel can carry on their normal jobs with greater confidence that they are performing in a safe and legal manner.

While not part of the pre-audit phase, information gathered during the audit must always be analyzed to improve future audits. Needless to say, the draft audit report must be made available to all members of the audited establishment for corrections, comments before the final report is published. In DND this also means availability to Environment Canada and other key government departments. The final report and its recommendations will be sent to all major participants in the audit process as well as made available to the public if they request it.

Appendix: Environmental Audits in National Defence

Audit Report Outline

1. Executive Summary
2. Background
3. Setting
4. Audit Findings
   (*a*) Environmental management:
      Activity/operation
      - Regulatory compliance
      - Policy and code of practice compliance
      - General observations

   (*b*) Air, and so on
5. Recommendations
   (*a*) Environmental/management
   (*b*) Air, and so on

Typical Functions Examined

1. Environmental management system
2. Air pollution
3. Water supply and water pollution control
4. Solid waste disposal
5. Hazardous materials management (including hazardous waste disposal)
6. Emergency response to spills
7. Occupational health and safety (WHMIS)
8. Natural (bio-physical) and cultural resources management (particularly in ranges and training areas)
9. Noise pollution
10. Fuel storage and handling
11. Bio-medical waste disposal

Objectives

1. To ensure compliance with federal legislation and regulations
2. To ensure compliance with internal policies and regulations
3. To ensure all non-regulated activities are carried out in an environmentally sound manner, in accordance with accepted codes of good practice
4. To ensure all activities are carried out in an environmentally ethical and moral manner (good corporate citizenship)
5. To ensure that the environmental management system is in place and being implemented to ensure compliance
6. To reassure senior management and the public that the Department is operating in a safe and legal manner

Pre-audit Steps

1. Organize/hire audit team
2. Review establishment organization and activities
3. Scope the audit
4. Develop detailed protocols or adapt generic protocols
5. Finalize audit team composition
6. Prepare pre-audit questionnaire
7. Make pre-audit visit
8. Review applicable regulations, policies, codes of practice
9. Review previous audit reports
10. Refine scope and audit protocols
11. Develop detailed agenda for the audit visit.

Outline of Pre-audit Procedures

1. Introduction
2. Policy and general procedures
3. Objectives of audit
4. Detailed pre-audit procedures
5. Concluding remarks.

Environmental Auditing Standards and Activities

D. Chollak

THE CONCEPT OF environmental auditing has evolved fairly rapidly over the past few years, yet it has not reached a mature state in Canada. That stage is still several years away, perhaps by the end of the 1990s. This paper will examine existing environmental auditing standards and a number of field techniques that are used to measure the performance of a facility or field operation.

The type of environmental audit to be conducted will determine which standards apply and what field techniques are to be used. Management audits will tend to rely on readily available information and likely will be undertaken in office settings. Exploration, production and maintenance or operational audits will rely on extensive field visits with limited review of available information in Corporate or Division office settings. On the other hand, baseline, compliance, acquisition and divestiture audits require extensive involvement at facility locations together with a detailed review of available information. The diversity of audit types currently available to auditors increases the need to establish environmental auditing standards.

Standards and Regulations

1. *Existing Standards.* Consultants, governments, and individual companies tend to use their own "in-house" audit programs. Generally, there are very few environmental auditing standards which can be used. Industry associations are now just developing their own "standards" although

D. Chollak is Corporate Environmental Advisor, Canadian Occidental Petroleum Ltd., Calgary.

individual company members are adopting and modifying these standards to suit their own needs.

The Canadian Petroleum Association (CPA) issued their Environmental Audit Guidelines in 1989. These guidelines were preceded by the CPA's Environmental Code of Practice and a revised set of Environmental Operating Guidelines. The CPA Environmental Audit Guidelines utilize the Code and the Operating Guidelines as a base or reference.

Another industry association, the Petroleum Association for the Conservation of the Canadian Environment (PACE) has recently prepared a draft of their Environmental Audit Guidelines. Other industry associations have either begun work on such guidelines or are giving it serious consideration.

Regardless of which environmental audit guidelines are used, their use is limited by the amount of information they contain. Most guidelines are merely "protocols" or "how-to manuals." It is not possible to provide additional standards in these guidelines or such guidelines will become very large or multi-volume. On that basis, other standards are required in order to complete the environmental audit. There are two basic types of standards which can be utilized, they are either external or internal, to the audit guidelines.

2. *External Standards.* External standards generally apply to federal or provincial acts, regulations, or guidelines. Municipal by-laws also apply but to a lesser extent than federal or provincial acts or regulations. Depending upon the type of facility, the products it manufactures and where these products are shipped, U.S. and/or international regulations may apply.

In the case of oil and gas exploration and production fields, specific facilities or functional areas usually operate under the requirements of permits, licences, certificates or an equivalent approval mechanism. These "approvals-to-operate" clearly identify the limits for specific emissions, effluents and more recently, wastes.

External standards set the amounts by which facilities are either in, or out of, compliance. These standards are the most important criteria by which a facility is audited.

3. *Internal Standards.* Internal standards are generally set by a company to measure its own level of compliance against its own limits or requirements. There are a wide variety of internal standards which can be used. They include: corporate, division, district or plant-specific policies,

procedures, standards and guidelines. Industry associations and government agencies may have their own internal standards.

Industry association guidelines, codes of practice and "good operating practices" are also considered to be internal. The CPA and Canadian Chemical Producers Association (CCPA) and many other industry associations have recently issued their own industry-specific "standards." Member companies generally adopt and abide by these standards.

Audit Activities

Most environmental audits follow a standard set of procedures, namely: pre-audit activities and post-audit activities. This paper will concentrate on the audit activities as the pre- and post-activities are being addressed by others.

Audit activities will, in all probability, vary from company to company, between industries, or between government departments; however, the audit steps taken are usually quite similar. Key activities include a kick-off meeting, document review, initial facility tour, conducting interviews, preparing recommendations, final facility tour, and an exit meeting. All of these steps are critical to compiling, reviewing, and assessing the information by which a particular facility or operating area is to be audited.

1. *Kick-Off Meeting.* Kick-off meetings are usually held at the facility and are meant to act as an orientation for the audit team and key members of the facility.

It is not always possible to review the corporate and division office material at the facility due to limitations in space, manpower or time. There will obviously be documents that will only be found at the corporate or division office and those only at the facility. All documents will require review as part of the overall audit. The task of document review will be addressed in more detail below.

The formal kick-off meeting will include the audit team leader and all or most members of the audit team as well as the facility manager, assistant manager and the environmental staff. It is at the discretion of the facility manager to select other individuals to attend this orientation meeting.

The main purpose of this meeting is to establish grounds on which the audit is to take place. The following items are usually addressed:

- introduction of all personnel to be involved
- review the overall objectives of the audit
- explain how the audit results will be used and by whom
- review the key activities that will take place
- identify restricted areas at the facility
- obtain approval for the use of recording devices such as cameras
- review the tasks of the audit team members
- explain the level of involvement of the facility personnel and schedule interviews if not already established
- reconfirm the general audit schedule and adjust as necessary
- answer any initial questions or concerns.

Following the kick-off meeting, the formal activities of the audit can begin.

2. *Document Review.* Assuming the appropriate documents have already been reviewed at the various corporate or division office locations, the facility files will also require review as the same documents are usually not found at both locations.

The first step is to determine where these documents are located and who is in charge of maintaining them. Most of the documents will usually be found in one general (albeit not always a central) location, while others will be found in other locations scattered throughout the facility (i.e., in staff offices, the lab, maintenance shops, etc.). It is usually suggested that all documents be compiled in a central location as a pre-audit activity.

Reviewing the documents is time-consuming. But, if the audit objectives are kept in mind and the subject matter is carefully scrutinized, valuable information and a number of "surprises" will usually be found. The use of a document checklist is very helpful.

Depending upon the scope of the audit, a checklist of the following types of documents could be utilized:

- administrative (i.e., organization, training, internal reports)
- regulatory (i.e., acts, regulations, permits, correspondence, applications and inspections)
- technical (i.e., general facility information, water, air, wastes, storage, emergency response, and PCBs).

The checklist could also be used to evaluate the availability of the documents and the quality of such documents. The use of and review of questionnaires is optional.

A list of key issues, follow-up questions, and possible recommendations should be initiated. These lists should be discussed either during or after the document review period but it is very important that all issues get addressed in a timely manner. Failure to follow-up may lead to letting an important issue or possibly a recommendation "fall between the boards."

3. *Initial Facility Tour.* Following review of the available documents of the facility, the audit team will be better prepared to tour the facility and to make reliable observations of the various processes and environmental concerns.

Audit teams are generally not familiar with the day-to-day activities at the facility they are auditing and will therefore require a knowledgeable facility representative to explain the details of the processes and how certain environmental matters are dealt with (i.e., wastes, storage and sampling, etc.).

The team should attempt to follow the process streams from where the raw materials enter the facility, through the various processes, and through to the storage and distribution of the finished products. Particular attention should be paid to the waste streams such as emissions, effluents and solid or semi-solid wastes. Waste streams are more likely to be of concern than are process streams. Other areas of the facility should also be looked at such as: stores, maintenance, power plants, steam plants, specialized fabrication shops like fibreglass shops, etc.

Sophisticated plant tours will utilize tour checklists and/or other types of auditor aids, although the auditing experience of the team leader and the team members is what really counts. Audit experience allows the team to know what and where to look at the facility. Additional information can be obtained by asking questions of the facility representative. Notes should be taken of significant issues and flagged as possible recommendations. Questions that can't be answered by the tour leader should also be noted and then directed to the appropriate individual following the tour. Additional follow-up activities may arise from the tour, such as revisiting other parts of the facility, reviewing previously reviewed documents or talking to the facility manager and other personnel. Testing of wastes, effluents or air emissions is not usually done during the audit. However, if testing is warranted, a recommendation for sampling or testing should be made.

Documenting what has been seen can be enhanced through the use of cameras, videos or similar devices. However, use of these devices is optional and permission to use them rests solely with the facility manager. Approval must be obtained in advance of the audit. Safety considerations may also limit the use of cameras to specific areas of the facility.

4. *Interviews.* Interviews with key personnel are normally scheduled in advance, as a pre-audit activity; however some adjustments to the schedule may be necessary owing to shift changes, illnesses and/or operational problems at the facility. Attempts should be made to interview a broad range of personnel from all areas of the facility, including the facility manager, environmental representative or liaison staff, the facility or process engineer/supervisor, receiving/shipping department personnel, warehousing/stores personnel, and laboratory staff and secretaries who type and/or file environmental correspondence.

The purpose of the interviews is to obtain technical, operational and functional information from individuals that work in specialized areas of the facility and to identify potential environmental concerns or issues not readily available from other facility personnel or audit activities.

A major task of the interviewer is to relay the fact that the interview is confidential and that issues raised will be noted for follow-up but more importantly, that no personal liability is associated with the information provided by the interviewee. Information obtained from these interviews can lead to follow-up activities such as revisiting site-specific areas of the facility, other interviews, or a recommendation.

By the time all of the interviews are conducted and the appropriate follow-up activities have been completed, the audit team will be ready to formulate or develop their initial set of recommendations.

5. *Preparing Recommendations.* Recommendations arising from an environmental audit will:

- keep the facility in compliance,
- improve the administrative procedures that deal with the environmental matters,
- direct the facility towards meeting the requirements of industry association "standards," good operating practices, and
- increase the level of environmental awareness at the facility.

Preparation of the recommendations should be done by the entire audit team, in a remote location, away from the facility. Each member of the team must provide their individual observations of the facility and propose what they feel are required recommendations. The recommendations should be short and concise and may be grouped into specific areas of concern such as administrative, awareness/training, regulatory operational, etc. Prioritizing the recommendations is optional. All recommendations and significant observations should be presented at the exit meeting.

Certain issues and concerns identified by the team may not warrant specific recommendations but should be mentioned at the exit meeting and clearly explained in the written report and/or covering letter.

The entire audit team should be satisfied that from what they have seen, heard and understand of the facility they are auditing, that reasonable recommendations have been made. If the overall understanding is not clear to everyone, then additional information should be obtained as quickly as possible in order to finalize the recommendations. It may be necessary to revisit the facility, interview additional personnel or review documents.

6. *Final Facility Tour.* Uncertainties arising from the audit, particularly those raised during the development of the recommendations, should be investigated by the team members, identifying those uncertainties. It may be necessary to revisit a specific area of the facility, to ask additional questions of certain individuals or the re-review of a document in the files. This follow-up may be necessary to finalize a proposed recommendation or to develop additional recommendations from significant concerns or observations. All recommendations must be presented at the exit meeting.

7. *Exit Meeting.* The exit meeting is chaired by the audit team leader and attended by the entire audit team, the facility manager, the environmental representative and others the facility manager deems necessary to attend. The order and manner by which the audit information and recommendations are presented is very important at the exit meeting. An overview of the audit team's activities and the involvement of facility personnel is reviewed and acknowledgements are given.

A short discussion of the general environmental conditions of the facility is presented noting major accomplishments and work in progress. Less favourable issues are reserved for the discussion of the recommendations.

A list of the recommendations should then be distributed to all in attendance. The team leader should review each recommendation and then present the rationale for each one of them. A general discussion usually ensues with each recommendation.

Follow-up activities may include the facility management responding to the recommendations presented, and the audit team preparing a written report. Completion and/or implementation dates will likely be assigned to each of the recommendations. Adequate follow-up is also a very important aspect of performing environmental audits. Details of the various follow-up activities will be presented by other speakers.

Conclusion

The art of environmental auditing in Canada has come a long way in recent years but much remains to be done. Auditing standards are few, but many are in the developmental stage by both industry associations and by individual companies or government departments. Until auditing standards are well developed, internal and external environmental standards will continue to be used. These standards will always supplement environmental auditing standards. Different types of environmental audits require varying amounts of information that has to be reviewed and assimilated as part of the audit process. The seven fundamental steps to audit activities discussed above may not vary from audit type to audit type. The level of effort required to complete any type of audit is generally quite significant and may only be limited by the experience and resources available.

There are no simple ways to minimize the effort or the amount of information that has to be collected and reviewed during audits but, as environmental auditing standards are refined, and experience is acquired, the "art" of doing environmental audits will become easier.

The collective knowledge, experience and interest shown by the speakers and participants at this symposium is a true reflection of where environmental auditing is heading in Canada.

Environmental Auditing at BF Goodrich

T.J. Fletcher

BEFORE ADDRESSING the field work during environmental audits, I want you to understand that BF Goodrich is no longer a tire company. Although rubber and tires are an important part of the history of this company, chemicals, aerospace and defense have been playing an important role in the company for many years. In 1989 the tire business was sold leaving BF Goodrich with three divisions: Geon Vinyl, Specialty Polymers and Chemicals and Aerospace and Defense. There are several locations in Canada:

- a vinyl resin plant at Scotford, Alberta, northeast of Edmonton
- vinyl resin and compounding plants in Shawinigan, Quebec, and Niagara, Ontario
- a sales office in Montreal, Quebec
- a Canadian head office in Waterloo, Ontario.

Auditing Background

BF Goodrich has been auditing its facilities worldwide since 1978. The original audits were strictly detailed compliance audits. These audits were assurance to upper levels of management that all locations were not missing anything which could lead to problems with their respective authorities. They evolved after a few years because we were not finding any major issues.

T.J. Fletcher is Manager, Safety, Health and Environmental Laboratory, B.F. Goodrich Canada Inc., Fort Saskatchewan, Alberta.

Presently we are doing audits at all locations approximately every two years. The objective of auditing is to ensure that we are practising good environmental management. This point is very important: it means that simply maintaining compliance is not good enough. We can only minimize our impact on the environment and our future liabilities by doing more than we are asked for.

Pre-audit Preparation

A typical audit at one of our locations takes four to five days with an audit team of one or two people. We make a point of scheduling the audit so that the environmental staff and most of the management are on site for the week. This is necessary because these are key people if we want to examine environmental management.

An audit can be a tremendous learning experience for the auditor as well as the plant. You can use this opportunity to assign a new or junior person to assist someone who is experienced in auditing. In the case of a larger plant site it may be necessary to assign two experienced people to do that audit.

Before the appointed week the auditors will need to review past audit reports. It will also be necessary to familiarize themselves with the process and the limits placed on emissions. The audit team will need to be familiar with internal company standards. These would include:

- environmental policy
- environmental operating practices or procedures
- engineering standards.

Other areas the auditors will need to understand are association "Codes of Practice" or standards. The Canadian Chemical Producers Association (CCPA), for example, have developed six Codes of Practice. All codes have strong links with environmental issues and management. Many other industrial associations have, or are in the process of developing, similar codes and standards.

At this point in time the auditors may be given some direction by upper management to ensure that they audit a particular facet of environmental management. If there happens to be an issue on the horizon such as the reduction of or replacement of CFCs they may want to pay particular attention to it in the audit.

The Week of the Audit

The first few hours are taken up with an orientation to the site safety procedures and an introduction to the plant environmental staff and manager. At this point some general discussion amongst the auditors and the plant environmental staff is helpful for scheduling. If there are particular processes or procedures that only run periodically you can schedule the auditors to be ready at those times. Typically a BF Goodrich auditor will request interviews with a cross section of plant personnel including managers, supervisors, operating technicians and maintenance technicians. Since the plant is operating, these times will have to be scheduled in. Finally, an eye to the weather forecast is advisable. We spend approximately half of our time outside around the site. An auditor from Texas or Kentucky would appreciate not tramping through snow if possible.

Usually the first audit team is a review of outstanding audit recommendations. This is easy to do and helps establish relationships amongst people who will be spending a lot of time together. The auditor will highlight a few or all items to physically inspect during the week to confirm they have been resolved.

By this time half to three quarters of a day has passed. We will usually start the site inspection at this point. Each day is concluded by a meeting of the audit team and the environmental staff. This serves several purposes. The auditors review their findings and recommendations. The plant staff have the opportunity to correct mistakes or misimpressions. This is crucial. If an auditor makes a recommendation regarding the plant based on a mistake (e.g., wrongly identified piece of equipment or incorrect application of company or government standard) the credibility of the whole audit may be questioned. This daily review is a tremendous opportunity for the two groups of people to share ideas and learn. BF Goodrich often has a plant environmental engineer on the audit teams. Other plants may have resolved similar problems to the ones identified in the audit. Conversely, the auditors may see something positive that they can take back to their own plant.

By the second day we try to get into the meat of the audit. Most of the time is spent talking to people and site inspection. As mentioned earlier, the auditor will generally request interviews with several people. These interviews are done in confidence with no other plant people in attendance. Interviews do not have a formal checklist or protocol. Instead they are used to get a general impression of attitudes and perceptions. If people at the supervisory level think that cost or production is absolutely the most important then your operators will never place environment high in their priorities. They can shed light on the management systems. In order for a process to be effective it must have:

- written procedures
- people trained
- follow-up or auditing.

Talking to operators, supervisors and managers may reveal weaknesses in one or more of the three areas. The auditor can get a feel for the practices. We all know that equipment is often operated differently from how it was designed. The practice may have changed or evolved over years so that the change never became apparent. Finally the auditor may hear "confessions." Behind closed doors, plant people may feel more comfortable revealing their concerns regarding equipment design, operating procedures, plant practices or even perceived attitudes.

Field Work

Several aspects of environmental management are considered during the physical inspections:

- air, water and soil spill protection
- fugitive emissions
- waste reduction.

At this point in time we are not looking for major design flaws or deficiencies. These would be discovered during a hazard and operability review (HAZOP). If a HAZOP has not been conducted on an important piece of equipment or a modification, a recommendation might be made to conduct one. Often the items which cause environmental problems are modifications to a piece of equipment.

The auditor needs to be something of a detective to find these. A modification may appear as freshly painted piping amongst older equipment. A shiny new motor or pump may be simply a replacement or it may indicate a modification has taken place.

We think of spill protection and think of major systems and dollars to prevent it. This is not always necessary. While doing the inspection during an audit, look at the floor and walls. Stains may have been caused by something leaking or spraying. Consider raw materials. Many come in drums or returnable containers. Could a forklift driving by pierce one?

If it did leak where would the contents go? Maybe the sewer isn't the best place. Are incompatible raw materials stored together?

An area of spill protection which we are addressing more closely is the soil. The company policy is to eliminate underground storage tanks. Other storage areas outdoors should be inspected. Look for telltale stains. Are these areas paved and diked? Are there ways to minimize quantities stored?

Fugitive emissions has not been closely regulated in much of the petro-chemical industry. Because of the health concerns over vinyl chloride the vinyl industry has done a lot to minimize fugitive emissions. During the inspection look at relief valves. Is there any indication one has relieved? Are there rupture disks below a safety valve? Are double mechanical seals used on rotating equipment? There should be a schedule for leak checks with portable analyzers. How often are analyzers calibrated? What is the plant definition of a leak? What is the action taken when a leak is detected? How quickly is the follow-up taken?

Waste reduction is an issue most of us are concerned about. Landfills are getting full and expensive, and the potential liabilities in future are frightening. Questions to ask could be:

- how much do you generate?
- do you segregate it?
- have you examined alternative packaging with your suppliers?
- can you buy it bulk?

On the plant tour look into waste bins. This will help confirm answers to the previous questions and may be revealing. As you work your way through the plant you will find that the plant environmental staff will identify many if not most things. There may be items they simply were unsure about. There may be items they have not been able to convince management about. This is a legitimate and useful source of information.

Wrap Up

Throughout the week of the audit there will have been discussions regarding findings and recommendations. Generally we categorize recommendations into three types: The highest priority items are requirements. They have this priority because they relate to legislation, company standards or pose an immediate threat to the environment. Recommendations are items which generally must be followed but have some flexibility.

For example, a site could decide not to follow a recommendation if the estimate for resolving it cost too much.

Finally, housekeeping items are listed. These are minor sorts of things that are easily agreed upon and corrected. This list doesn't require wide distribution. In fact many are usually corrected before the audit is complete.

A wrap-up meeting is always scheduled on the last day. Plant management attends along with at least the Vice-President of Manufacturing. The audit team presents its findings even though the formal report is not complete. This meeting is usually short (i.e., one to two hours). The audit requirements and recommendations are presented and the plant manager commits to implementing them.

The final written report is sent to the plant manager and the environmental staff two to three weeks later. The plant has ninety days to respond. Generally, the only accepted responses are: complete or planning to be completed by a given date. As noted earlier, the items listed as recommendations may be excused given sufficient justification. The teeth to these audits lie in the fact that resolution of the items will influence performance ratings and salary. I am not aware of any problems accepting recommendations because environmental audits within BF Goodrich are considered "win-win" endeavours. Good environmental management, like safety and health, means less hassle and liability in the future.

Environmental Audits: A Practical Perspective

C. Pelletier

SINCE THE U.S. congressional passage of the National Environmental Policy Act, in 1969, both the U.S. and Canada have witnessed a proliferation of environmental laws and regulations. They continue to accumulate at an unrelenting rate. Not only are there considerably more laws and regulations to violate, but the penalties for so doing are becoming increasingly severe. Indeed, with the enactment of the Canadian Environmental Protection Act, government is empowered with an even more comprehensive and enforceable statute with which to prosecute corporations in contravention of environmental regulations.

The environmental audit, while yet in its infancy, has evolved as a logical response by management to an ever-increasing list of environmental liabilities, which often have the potential to result in plant shutdowns, prosecution, and/or major financial penalties. By identifying noncompliance and environmental liabilities, the environmental audit is a fundamental component of a comprehensive defensive strategy for corporate survival in an environment of increasing regulation and litigation. It is the purpose of this paper to encourage environmental audits and discuss the essentials of their practical application.

What is an Environmental Audit?

While the comparison is sometimes made to financial audit, because environmental auditing is still in its infancy, a well-accepted definition

C. Pelletier is Chief of the National Capital Commission, Ottawa.

of the concept does not exist. As used in this paper and as generally understood, environmental auditing may best be described as follows:

Environmental auditing is defined as the use of a systematic, documented, periodic and objective review by entities of facility operations and practices related to meeting environmental requirements whether such facilities are publicly or privately operated. Audits can be designed to accomplish any or all of the following: verify compliance with environmental regulatory requirements, evaluate the effectiveness of environmental management systems already in place, or assess risks from regulated or unregulated materials and practices. In this way, environmental auditing can be an effective tool in an overall environmental management program.[1]

Continuing the parallel to the annual financial audit where the financial statements are prepared by management, and then reviewed, tested and reported on to the shareholders by the outside auditors, the environmental audit is perhaps more akin to the financial auditor's practice of providing a management letter commenting at some length on areas in which there is room for improvement.

As described, an environmental audit is a management tool. It leads to an internal document. To the knowledge of the author, environmental audits are not yet mandatory in any jurisdiction. Perhaps this accounts for the absence of an accepted definition.

The United States Environmental Protection Agency has, however, published an Environmental Auditing Policy Statement.[2] In this statement the EPA adopted a definition of environmental auditing which is virtually identical to that quoted above. The EPA very specifically encouraged regulated entities to "institute environmental auditing programs to help insure the adequacy of internal systems to achieve, maintain and monitor compliance."[3] In its Draft Enforcement and Compliance Policy issued for discussion in the Spring of 1987, Environment Canada also proposed two environmental audits.[4]

Why Do an Audit?

In the absence of regulatory requirements, why are companies undertaking environmental audits? The answer is not to be found in the words of encouragement from government. Instead, it lies in the growing realization, born out of self-interest, that it is good business practice for management to have such an evaluation undertaken.

The self-interest and good business practice are rooted in the potential exposure of the corporation, and, to a lesser extent, its directors, officers and employees to prosecution, penalties or civil liability if environmental problems are not identified early and properly managed.

The environmental audit may also be said to be due diligence in action. Generally speaking, offences under environmental statues have been classified as strict liability offences.[5] One of the few defences available to an accused who is charged with such an offence is to show that the offence was committed without his knowledge or consent and that he exercised all due diligence to prevent its commission.

It is difficult to imagine a more responsible approach for the board or senior management of a company to take than to commission an environmental audit that actually examines in a systematic and professional manner the corporation's record of compliance, its system of internal management and control, and attempts to identify possible future problem areas. Whether it will constitute due diligence may depend largely on what happens to the audit report after it is reviewed. It is certainly not going to constitute "all due diligence" as the report's recommendations would have prevented a spill if only they had been implemented. On the other hand, if a professional report is received and its recommendations are implemented, from a practical point of view, there is little else that directors and senior management can do. It would be going too far to suggest that by implementing the audit recommendations, the defence of all due diligence could be established. While it would, without question, be viewed favourably, there is no case of which the author is aware where the court has been asked to determine whether an audit can form the basis of a due diligence defence. While it is submitted that to require any higher standard would be to impose an unrealistically high burden on corporations and the people who run them, it is likely that the courts will use the wisdom of hindsight to carefully scrutinize the environmental audit to determine whether the offence could have been prevented by a more careful audit.

In addition to the operational audit, environmental audits are also finding their way into an increasing number of commercial transactions.[6] They can be part of the purchaser's due diligence or may be a precondition to a lender's loan advance.[7] In some cases, audits may be required to ensure that securities regulatory disclosure requirements are discharged.[8] Finally, and most strikingly, in some parts of the United States state law requires operators and owners of commercial or industrial businesses to undertake an environmental audit and obtain an environmental "clean bill of health" before certain kinds of commercial transactions can be completed, and audits are being used for this purpose.[9]

In British Columbia, it is likely that we will increasingly see municipal planners and development control officers asking for audits before they will issue development permits for industrial lands. This requirement will be even greater where the lands will move from industrial to residential to commercial zoning.

Environmental Audit Functions

In essence, the environmental audit is a verification of the existence and competent application of appropriate environmental management procedures. The audit will generally operate within a defined regulatory framework and focus on worker health, safety and the external environment.

More specifically, the means by which an audit is conducted will depend upon its intended purpose, which may include any of the following:

- gauging the extent of potential environmental liability associated with day-to-day operations of a company or its products:
- identifying regulatory infractions which may lead to prosecutions;
- evaluating the adequacy of procedural or management practices that may lead to environmental, health or safety concerns;
- critically reviewing the costs of compliance with environmental laws; and
- estimating the degree of environmental risk in property acquisitions or dispositions.

Audit Personnel

Proper selection of audit personnel is extremely important as the company is at great risk if the audit recommendations are inappropriate and are consequently ignored. Similarly, audit omissions may expose the company to unknown liabilities. Ideally, a combination of technical and legal skills are required. Previous experience in planning and implementing environmental audits is an essential qualification of any audit team. Due to the requirement of objectivity, internal company personnel must normally be excluded from the audit team. However, personnel from one company division may sometimes be employed to audit another division. The cost of completing an audit, or the potentially crippling liabilities of failing to do so properly, enforces an important principle: choose your auditor very carefully.

Conducting an Audit

A typical minesite environmental audit will generally require three to five days in the field and an equal amount of time for write-up of the audit report. This represents a nominal amount of effort when balanced against the benefits provided to the corporation upon the audit's completion. While the time required for conducting an audit will vary with its scope and objectives, it is difficult to imagine a circumstance when so doing would not be a cost-effective exercise. The elements of the audit, in flow-sheet format, are shown in Figure 1.

Defining the scope and objectives. An audit will commence after the auditors and company have reached consensus on the objective and scope of the auditing program. Audit objectives will include the following:

- the technical and regulatory scope (encompassing any or all of health, safety, and external environment);
- the depth of the assessment (review of files, correspondence, monitoring data, or actual on-site monitoring of discharges by the auditors, site inspection, etc.);
- means of attaining the desired degree of confidentiality;
- means of assuring top level management support; and
- timing and budget.

Pre-audit Checklist. A comprehensive pre-audit checklist should be prepared and provided to the company as a means of identifying topics to be covered by the audit. These should include persons to be interviewed, files, records and permits/licences to be reviewed, and plant operations to be inspected. This preview will enable management to properly prepare for the audit, and have on hand the necessary reference materials and schedule for interviews. Many auditors recommend the preparation of a detailed questionnaire rather than a checklist, but it is the author's experience that this inhibits spontaneity and unanticipated lines of questioning. Typically, the checklist will be thirty or forty pages in length, allowing space for the auditor to complete notes based on plant inspections, interviews and records review.

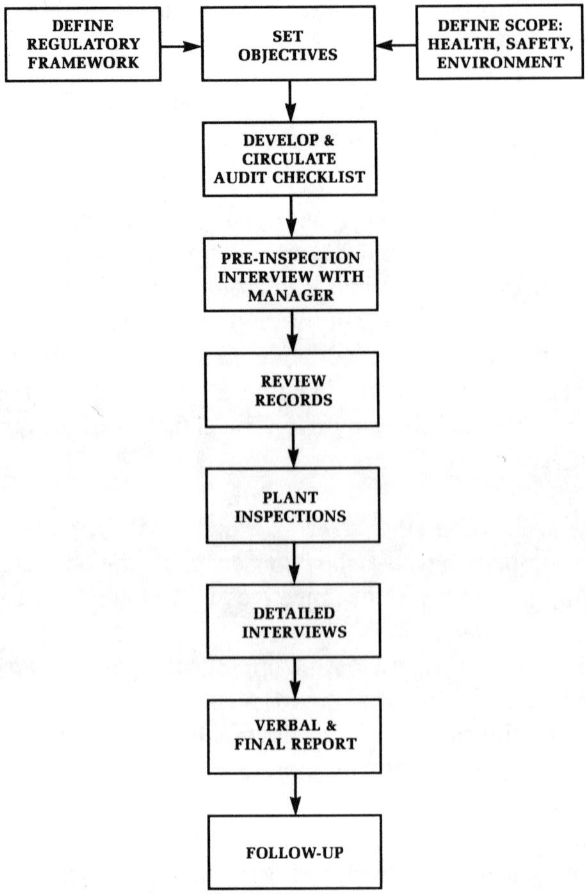

FIGURE 1. ELEMENTS OF AN AUDIT

Introductory Meeting. If the audit is the first of a series, it should commence with an interview with the General Manager to obtain an overview of the operation's history, its compliance record, public/government environmental issues, spill occurrences, and known liability concerns. At this meeting, the auditors will outline the procedure and timetable for the audit and the mechanism for reporting their findings.

Inspection of Records. With all of the pertinent files and records set aside in advance, the auditors will carefully review all correspondence with regulatory agencies, permits/licences, monitoring data, environmental reports and files, spill reports, contingency plans, tailings dam inspection reports and all matters relating to environmental management. This is one of the more time-consuming, but also one of the most important tasks and, accordingly, should be given high priority. Many revealing aspects of compliance history and potential liability will be revealed by this exercise, and items will be noted for follow-up during interviews with plant personnel.

Plant/Mine Inspections. Inspections are then carried out in collaboration with the various senior plant or mine personnel or other members of management who are intimately familiar with all aspects of the operations and the compliance issues. Detailed notes should be kept and photographs taken. The usual audit will not include sampling or on-site analysis.

Interviews. Persons to be interviewed should generally be chosen in advance and should consist of those responsible for the following (as appropriate):

- environmental policy and procedure;
- permit and licence compliance;
- effluent, air emissions and solid wastes monitoring;
- safety and hygiene;
- surface reclamation;
- tailings pond construction and operation;
- operation of waste dumps;
- disposal of waste oils, PCBs and other hazardous wastes;
- spills prevention and response;
- warehousing and laboratories; and
- fuels and explosives storage.

Verbal Report. At the conclusion of the audit visit, the auditors should meet with the General Manager, the Mine and Mill Superintendents, and the Environmental Supervisor and present their findings verbally. This briefing should allow for clarification and discussion of any recommendations before they are written into the report.

Written Report. This report is the final summary of the auditors' findings and recommendations and, as such, must be extremely well thought out and expressed. The written report usually takes as much time to prepare as the field inspection itself.

Elements of an Effective Audit Program

The above generic approach to conduction an audit will fulfill the needs of most operations that are concerned about their compliance status and potential environmental labilities. Many variations to this approach are possible, depending upon the auditors' style and experience, the audited company's management policy with respect to environmental affairs, and the specific objectives to be achieved. However, regardless of circumstance, an effective environmental audit must incorporate the following critical elements:

- top management support for environmental auditing and commitment to follow-up on audit findings;
- adequately trained auditors who are objective and independent of the audited function;
- defined audit objectives and a formal approach for carrying out the audit;
- an audit report which clearly states the objectives, findings and required corrective actions;
- means of assuring quality and completeness of the audit; and
- follow-up to ensure that the recommended corrective actions have been completed.

Conclusion

To conclude, it is worthwhile to review the benefits which will accrue to a corporation from commissioning an environmental audit versus the possible consequences of choosing not to.

The overwhelming benefit of an environmental audit is the increased awareness of vulnerabilities which will come with an independent evaluation of environmental systems and compliance. Provided that the audit is taken seriously by management and its recommendations followed, the company will react to the audit by improving measures for attaining permit compliance, minimizing safety, health and environmental hazards and planning for emergencies. Ideally, the result will be an operation fully in compliance with the law, awareness of its vulnerabilities, and prepared to meet any environmental emergencies which may arise.[10]

As a result, the company will be less likely to be cited for regulatory violations and will pay fewer fines. Its corporate and public image will be enhanced. Directors and managerial personnel will rest more comfortably in the knowledge that their environmental policies and compliance efforts are less likely to contain serious flaws. Should the company become involved in a lawsuit in spite of these efforts, it will be able to demonstrate that having the audit carried out was a reflection of a policy of environmental conscientiousness which will assist in its defence and reduce potential penalties.

Notes

1. Arthur D. Little Inc., *Current Practices in Environment Auditing.* (February 1984). Janette M.F. MacDonald, *Environmental Auditing* (CBAO Conference, "Corporate Environmental Responsibility," 22 May 1987).
2. Federal Register, Vol. 51, No. 131, Wednesday, 9 July 1986, 25004.
3. Ibid., 25004.
4. Draft Enforcement and Compliance Policy, Environment Canada, Spring 1987 at p. 5. Statement of the Honourable Tom MacMillan as reported in Blair S. Taylor, "Environmental Audits: The Legal Technical and Practical Issues," paper presented to the Canadian Institute Environment Law Seminar in Toronto, 2 February 1988.
5. The case which identified environmental offences as strict liability offences is *R.* v. *Sault Ste. Marie* (1978) 40 C.C.C. (2d) 353 (S.C.C.).
6. See Charlotte L. Neitzel, "Environmental Risks and Liabilities in the Purchase and Sale of Mining and Milling Properties" (1985) 31 *Rocky Mountain Mineral Institute* 10–1.
7. Scott Wilsdon, "When a Security Becomes a Liability: Claims Against Lenders in Hazardous Waste Cleanup" (1978) 38 *Hastings Law Journal* 126.
8. Kenneth R. Myers and Thomas F. McCaffery III, "The Goals and Techniques of Environmental Audits" (1984) 30 *Practical Lawyer* 15.
9. David B. Farer, "Protecting Business Clients Against Prospective Environmental Legislation" (1987) 33 *Practical Lawyer* 15.
10. See J.B. Brodie and C.A. Pelletier, *The Role of Auditing in Demonstrating Environmental Due Diligence* (Vancouver: Rescan Environmental Services Ltd., 1988).

Post-audits: A Consultant's Perspective

H.P. Cleghorn

THE PREVIOUS SPEAKERS dealt with what have become accepted as the key elements of the environmental auditing strategy. Specifically, these were:

- pre-auditing;
- site investigation; and
- additional sampling.

The establishment of strict protocols, including checklists, interviews, site observations and sample data, are fundamental for the successful completion of these audits, which would lead to clear conclusions and an associated action plan. During this session, Dr. Boucher and I will cover the post-auditing phase of the audit, during which the available information is evaluated in terms of facility compliance with existing legislation, the management programs in place for dealing with these issues and the management reporting structure.

Although there is no specified procedure for the post-audit process, certain guidelines are routinely followed by most external consultants in this phase. These are:

- debriefing of facility personnel;
- report preparation and submission of a "draft" copy;
- report evaluation by legal counsel and development of an action plan; and
- follow-up procedures.

H.P. Cleghorn is Principal Consultant, Concord Scientific Corporation, Toronto.

In this presentation, I shall be discussing items 1, 2 and part of 3 (legal counsel evaluation): Dr. Boucher will cover the remaining issues and provide a summation of the overall post-auditing process.

Debriefing

In practice, two debriefing sessions are held. The first takes place on-site at the completion of the audits with the facility environmental coordinator, unless this person is already a member of the audit team, and the facility manager. During this session, an oral presentation of the observations, results and conclusions, if any, is made. The purpose of this debriefing session is twofold: first, it is a means of showing the team's courtesy and appreciation for what is always a disruption of the facility's operation—no matter how small; and, secondly, it sets the stage in an in-hostile atmosphere for presenting those regulatory non-compliance issues identified in the facility and the audit team's evaluation of the management structure and reporting. This debriefing session is also important because areas of non-compliance identified in the audit that require immediate attention are quickly brought to the attention of the facility manager.

Approximately one to two weeks after the audit, an audit memorandum is prepared by the team for distribution to the facility environmental coordinator and the facility manager for review. In certain cases, particularly for acquisitions and divestitures, this memorandum may also be circulated to legal counsel for information purposes only. The information included in this memorandum may be incomplete, particularly if additional on-site sampling and analysis were required during the audit; however, it provides a written account of the key issues presented at the debriefing session.

The format for the memorandum (no more than four to six pages) is optional; however, the one used by Concord is usually covered under the following headings. All pages are stamped "Draft. For Review Only."

Title:	Summary Field Notes for Environmental Audit of Facility "*X*"
Audit Dates:	12–15 December 1989
Auditor:	J.P. Smith, P. Eng.

Background Interviews

Personnel Interviewed:
Site History:
Process Description:

Management and Training

Management Structure:
Management Programs:
Senior Staff Responsibilities:
Emergency Response:
Training Programs:

Regulatory Issues

Air Related Issues:
Water Related Issues:
Waste:

Construction, Storage, Handling and Shipping

Storage Facilities:
Receiving and Shipping:
Construction Materials:

Maintenance

Maintenance Programs:
Confined Spaces:
Pesticide and Herbicide Use:

Conclusions

Recommendations

Additional Sampling:
Emergency Non-compliance Issues:

Audit Report

After a total elapsed time of approximately four weeks from the site investigation, a "draft" copy of the audit report is submitted by the audit team leader for review by facility and corporate management and by legal counsel. Numbered copies of this draft report are prepared to limit circulation and maintain security. This document is reviewed in detail by legal counsel to ensure that there are no statements, inferences or conclusions that may be misinterpreted in terms of the facility's compliance with regard to existing or proposed new legislation. In addition, recommendations are reviewed by legal counsel with regard to the facility's and corporate manager's liability and, where applicable, the risk incurred by the corporate manager and the facility as well as the report's impact on any financial instruments held by third parties and facility insurability.

The important elements of a good audit report are (Cahill and Kane, 1987):

Technical

- Attend to details.
- Be thorough and precise.

Presentation

- Establish a consistent reporting format.
- Use clear and concise language. Avoid sensational language.

Security

- Have report reviewed by legal counsel.
- Number reports and limit distribution.

Attention to Details

- In this respect, an environmental audit is exactly the same as a financial audit. If the information is incorrect, vague or not supported by facts, then the credibility of the entire document is put in question.

Precision and Thoroughness

- One of the goals of the environmental audit is to provide the means for developing a follow-up plan. Consequently, once those areas of regulatory non-compliance, poor internal reporting or ineffective environmental management programs are identified, the facility must be able to quickly determine the scope of the problem and take the necessary corrective action. To take a simple example, if thirty-four waste manifests are examined and entries are incorrectly entered, then the number of manifests, the type of error and the required corrective action must be noted in the audit report.

Omission of important information which thereby renders the facility and its corporate manager liable is one of the worst professional failings and consequently the nightmare of every good auditor. No one likes environmental surprises such as underground storage tanks of PCB-containing oils or buried asbestos on-site after the audit is completed and the report has been submitted! In such cases, the legal, financial and insurability impact on the facility is often disastrous, in today's environmental climate, particularly in Ontario and Alberta.

Reporting Style Format

- Although the format should be consistent, there will be minor variations depending on the environmental audit objectives. For simplicity, we will consider the following examples:

  Auditing for organizational effectiveness: Such audits are usually conducted as "self-audits" by employees of the corporation, although if it is a "first-time" comprehensive audit, it may be conducted by an external auditor. Typically, the recipients of the report will be corporate managers and directors and the corporate lawyer. Reporting may be terminated at the "Memorandum" stage followed by the development of an action plan and follow-up procedures. The formal report, if required, will be concise and pithy and will include an executive summary with all the key points that is a maximum of six to eight pages long.

  Auditing for acquisition and divestiture: Such audits are usually "base-line," are treated as "first-time" audits and are usually conducted by external auditors. The driving force for each audit is usually whether the facility is in compliance with provincial and federal legislation and municipal by-laws. Recipients of such reports include legal counsel for both parties, corporate managers and lawyers from the respective parties, as well as any financial institutions and insurers that are involved in the transaction.

  When acquisition and divestiture is the primary objective which initiates the environmental audit, then a formal report is required which includes the following:

- Executive summary—eight to ten pages maximum;
- Auditing methodology (objective, scope, regulatory framework and implementation);
- Audit findings (discussion, conclusions and recommendations); and
- Appendices (including pre-audit checklist[s], sampling and analysis, borehole logs, photographs, etc.).

Clarity in Reporting

- It is essential that information is expressed in a concise, unambiguous manner in the report. Despite the security (discussed below) attendant on the circulation of this report, a billion dollar transaction can be jeopardized by poor wording or incorrect emphasis. For instance, extensive use of asbestos material in plant construction should be noted in a designated substance hazard assessment, but it is only a major environmental problem if it is friable or poorly covered (e.g., in pipe insulation or subject to erosion by contact or otherwise in a heavily trafficked area). The manner in which this information is presented in the report is, therefore, very important in determining potential facility liability.

Legal Counsel Review and Distribution of Reports

- Whereas environmental auditing is not regulatory driven in Canada as it is in the United States; in many ways, the impact of the report is the same whether used as a management tool or as evidence to show "due diligence" in arguing against cases of "wilful" negligence. For this reason, all audit findings must be phrased in such a manner that there is no ambiguity or provide cause for legal misrepresentation.

  To ensure security, there appears to be a generally accepted policy for auditors to limit the distribution and to number all copies of both the "draft" and final reports that are circulated. The draft report copies are destroyed after agreement is reached on the final report.

  One of the best guides for use in report preparation is presented in Blakeslee and Grabowski (1985). The authors stress the need to use a consistent format in reporting and recommend that the report be organized into at least four sections: introduction; scope of work; review of the findings; and conclusions. They also recommend that audit reports intended for corporate management or those that are very long and detailed include an executive summary at the beginning. The type of information included in each section is presented in Table 1.

Proposed Audit Report

Although the report format is optional, the preceding guidelines may be used in developing the following "generic" format for the "draft" and "final" reports. This format is currently used by Concord Scientific Corporation and several consulting companies based in Ontario and the Maritimes.

- Title
- Audit location, auditors and audit dates
- Executive summary—Includes sufficient information regarding objective, scope, observations, conclusions and recommendations that reference to remaining sections of the report is only necessary if specific data are required for review.
- Audit methodology—Includes the objective, scope, legislative review, environmental management structure, reporting and responsibilities, training programs and their implementation. Also, environmental issues are examined, including site history, e.g., wetlands, changes over at least the last ten to twenty years of the terrain, proximity to water courses. Multimedia emissions, including Certificates of Approval for pollution control devices and monitoring programs. Sanitary and storm sewer discharges and the municipal sewage waste treatment plant. Environmental noise, light (and radiation issues from the facility and their impact on the community are also evaluated.
- Audit findings—The observations and conclusions are discussed in this section in the context of current management programs and procedures and regulatory requirements. Recommendations in terms of complying with existing and proposed new regulations are presented. In general, such recommendations are limited since they are not considered fundamental to the audit. It is agreed that the audit should provide management an overview of liability and risk. Recommendations and a follow-up plan take place in further consultation.
- Supporting information and data—This section is included in appendices attached to the report and may be every long (twenty to one hundred pages), depending on the scope. Such information covers the

TABLE 1. AUDIT REPORT ORGANIZATION

INTRODUCTION

- Objective
- Audit dates and participants
- Audit methodology
- * • Summary of mandatory recommendations for plant management

SCOPE OF WORK

- ** • Facility description and history
- ** • Environmental impact of facility
- Environmental C of As, date of issue, restrictions, etc.
- ** • Review of federal, provincial and municipal legislation and bylaws
- Review of corporate environmental policies
- List of notices of violation, control orders, fines, complaints, etc.

REVIEW OF RESULTS

- Multimedia discharges from the plant:
    - air pollution, water effluent, sewer discharge, solid waste depository
    - hazardous waste handling and disposal, oil storage, PCBs, etc.
- ** • Assessment of each item with regard to regulatory compliance, noting proper operation as well as deficiencies

CONCLUSIONS

- ** • Summary of observations
- * • Recommendations re: items needling prompt attention
- * • General recommendations re: (policies and procedures that would assure better compliance
- ** • Summary of facility's status, including liability assessment, and adequacy of environmental control systems to meet current and anticipated future needs
    - Summary of facility's reporting requirements to meet regulations and corporate policies

* Audits (e.g., self-audits) for facility management only. ** Audits for corporate management only.

Source: Blakeslee and Grabowski (1985).

pre-audit checklist(s) interviews, air photograph information, bore-hole logs, sampling and analysis, analytical data and illustrative photographs).

Report Evaluation by Legal Counsel and Action Plan

Following review of the draft document by legal counsel, the final report will be submitted which includes all pertinent changes or modifications. All copies of the draft report are returned to the audit team leader and these copies are destroyed. Numbered copies of the final report will be submitted to the original circulation.

A final debriefing session will be held once the final report has been reviewed by all key parties at which time the recommendations are discussed and an action plan is established to execute these recommendations. In addition, a timetable will be established for the following:

- implementation of recommendations; and
- timetable for follow-up audits.

Follow-Up. The publication of the final report, particularly for an initial audit, leads to the follow-up to correct any deficiencies brought to light by the audit and an action plan to prevent their reoccurrence. Indeed, many large companies, particularly those with branch offices, or whose parent company is located in the United States, have instituted a program of follow-up audits (normally "self-audits") to assess the effectiveness of any new "corrective" programs. Even where regulatory compliance is being met, these audits are essential because environmental regulations are in a constant state of change and plant operating change to meet new production quotas as well as change in personnel.

References

Blakeslee and Grabowski. (1985). *A Practical Guide to Plant Environmental Audits.* New York: Van Nostrand Reinhold Company.

Cahill and Kane. (1987). *Environmental Audits* (5th ed.) Maryland: Government Institutes, Inc.

Post-audits

G.L. Boucher

THE AUDIT REPORT(S) and action plan derived therefrom are the heart of the paper trail associated with the audit process. Without debating the need or the subject of the paper work of your audit program to a solicitor/client privilege, the documents derived from that program should be kept confidential and not used for public relations.

Safeguards to protect confidentiality need to be implemented (limited copies, restriction of access, policy on recordkeeping, etc.) because the information contained in your audit documents can be the source of self-incrimination. The safeguards should not suffocate the *required* internal flow of information and definitely control its external release.

The reporting and follow-up procedures of your environmental audit should further the main objectives of your environmental policies of which the audit program forms a part. The post-audit format you choose can help increase environmental consciousness at your operating facility. This normally means involving plant personnel in the audit process.

Reporting Audit Results

Purpose

Taking into account that the audit report (verbal or written) forms the basis for communicating the findings and activating, when necessary, an action plan to address such findings, the report should be written in a

G.L. Boucher is Manager, Environmental Compliance, Domtar Research Centre, Domtar Inc., Senneville, Quebec.

format understandable and useful to the persons responsible for the implementation of such an action plan.

The audit report, even if it includes recommendations, is a rather stagnant document when not translated into an action plan. The audit report is only *part* of the paper trail verifying "due-diligence." By itself, the audit report is a self-incriminating document that has not realized its purpose.

Content

The report should be a working document, not a tome. The findings are to reflect, in a clear, concise, complete and accurate manner, the facts purporting to the environmental non-compliance and risk issues. One must avoid any speculations, generalities and adjectives which can only mislead or confuse the persons who have the authority to act upon the report. It should simply identify and describe the environmental matters which form the basis of regulatory or non-regulatory compliance.

The findings ought to be divided by categories: atmospheric emissions, hazardous waste and regulated substances, liquid discharges, spill prevention and contingency measures and recordkeeping. The important aspect is to choose the categories associated with your plant's operations and to maintain these same titles and reference points throughout your audit program.

In order to further facilitate the write-up of the action plan, the report can be divided between regulatory and non-regulatory matters, the latter referring to environmental risks. Regulatory matters can then be easily identified and addressed on a "fast-track basis."

Without curtailing audit report improvements, the important element is to be consistent in the format you choose because there is no perfect way to write this report. Everyone involved in the audit process should be at ease working within the audit format. The procedure should not overshadow the objectives of the audit program and environmental policies of your company.

Recommendations

Depending on the purpose and goals of the audit program, some will incorporate recommendations in their audit report. The problems created by such inclusion are three-fold.

First, the recommendations have a habit of being considered as the ultimate solutions and, when not acted upon, oblige the persons responsible for implementing corrective measures to justify any action or lack of action relating to the recommendations. Further they add to the paper trail because written justification will be required in order to present any "due diligence" concept. This comes about because the recommendations are associated with specific audit findings which define regulatory and non-regulatory matters.

Second, you must satisfy yourself that those making the recommendations are technically qualified to render such recommendations. This includes clear understanding of the process, the equipment, the available products (raw materials and replacements ...) and associated costs, the end-products (quality control ...).

Third, those making the recommendations should be familiar with the corporate context in which they operate. By corporate context I refer to the decision process within the company, the plant's objectives and needs, the risk factors, the marketplace....

If recommendations are included in your audit report, they should be made in conjunction with plant personnel. There will be fewer circulating documents which contain corrective measures not yet implemented.

One can partially eliminate some of the concerns about including recommendations in the audit report by proceeding directly to a preliminary action plan of the measures required to address the audit findings. The goal is to have the plant manager and his designated personnel develop a preliminary action plan with the participation of the audit team. Even if the preliminary action plan does not generate a solution to each finding, this approach brings the objectives and the energy required to solve the problems into focus in a timely fashion. When the problem is complex and requires further study, the preliminary action plan need simply state the steps to be taken in order to conclude such study. Discussions are held concerning findings that are subject to further investigations; but such findings can still be mentioned in the preliminary action plan. Each item in the action plan is subject to a reasonable timetable.

The advantage of this format is that it is action-driven and involves everyone in the audit process (the plant manager being the central figure). It also emphasizes the importance of post-audit activities to those charged with these responsibilities. The report itself becomes, as it should, a secondary document. The action plan is the working document tailored to the needs of the plant, and to which they can more easily relate and re-adapt, if and when required. The plant personnel are the people who will implement corrective measures.

Distribution of the Report

The distribution of the report starts at the exit meeting. All the findings (including those requiring further investigations) should be verbally reported to the plant manager and his designated personnel. This will help build a relationship of trust and bring the plant personnel, at an early start, to understand and appreciate environmental concerns. Most importantly, the plant manager will have the opportunity to confirm, acknowledge or correct the findings.

The distribution of the written audit report (which can be segregated into two distinct reports: regulatory matters and environmental practices (environmental risks) can be realized by the following scenario:

1. The audit team prepares a "draft" report within two weeks of their exit meeting.
2. The draft report is circulated to the plant manager and legal department for comments and returned to the audit team leader within two weeks.
3. Based on their comments, a final report is completed by the audit team within two weeks of reception of said comments and sent to the legal department and to top management when the audit is not subject to a solicitor/client privilege claim.
4. The legal department prepares a legal opinion based on the report and distributes such opinion to top management.

Item 4 of this scenario is required if the audit performed is subject to a solicitor/client privilege. In this scenario, the request for an audit by the audit team would have to generate from the legal department. If you decide to follow this route, I suggest you consult legal counsel because there are many requirements to be fulfilled in order to attempt to claim solicitor/client privilege.

Nevertheless, even if the audit report is not subject to solicitor/client privilege, the audit report should have limited circulation and only those who have a need to know should be entitled to access the report.

The audit report should be kept at the plant only until it has served its purpose. This occurs when the plant has been able to forward its comments to the audit team, understanding that the comments are given with full comprehension of the report which has been translated into an action plan. Audit reports including recommendations may tend to remain "active" for a longer time.

Follow-Up: Action Plan

The follow-up process is the administration of the action plan until its completion and full implementation. The objective is the ability to present a document confirming that all required corrective measures have been implemented and solved the environmental concerns addressed in the action plan. Such a paper trail has the real making of a due diligence defence, especially when the corrections were performed on a timely basis.

The action plan must provide a timetable for the accomplishment of each corrective measure listed in the action plan. There is a tendency to underestimate the time required to accomplish the tasks. Unrealistic deadlines put undue pressure on the plant and are often self-defeating and may oblige some extra degree of accountability in a due diligence presentation. It is important to set a realistic timetable.

Nevertheless, in dealing with regulatory matters, these should be tackled on a "fast-track basis." The objective is to highlight regulatory matters and give priority in undertaking the corrective measures associated with them. On occasion, non-regulatory matters expose the plant to an important environmental risk that requires immediate attention.

The final action plan should be reviewed by the audit team in order to ascertain that all findings have been understood and addressed by the plant. There should be a pre-set time for the production of such a plan. Work on the action plan starts when the final audit report is finished, and the plan should be distributed at least to those who have the authority to approve and implement the required corrective measures.

A progress report should be expected every three or six months. The plant manager or his superior identifies the corrective measures implemented. The report will be on-going until all measures listed in the action plan have been addressed. The quality of the measures undertaken can be ascertained during the next audit.

Finally, there is the audit of the audit. This audit can be performed under the same audit principles whereas its objectives and scope are well defined and provide for the optimum impartiality. This will assure the proper evolution of your audit program.

Conclusion

Post-audit activities are a reflection of the genuine objectives of your audit program. The primary objectives are identifying and *correcting* environmental non-compliance matters and environmental risks occurring or likely to occur at your plant. (You may wish your audit to concentrate solely on compliance matters.) Any non-compliance matters discovered during an audit and not diligently corrected may subject the company and individuals to criminal liability unless there is a legal basis for not correcting the matter.

The post-audit activities, especially with regard to the implementation of the action plan, must be set-up in order to maximize the participation of those responsible for such implementation. It is possible to maintain the required objectivity and independence of an environmental audit when some member(s) of the audit team is (are) plant personnel from the audited plant or, preferably, personnel from a similar plant.

A plant's environmental consciousness can be evaluated in its interest and action pursuant to the action plan derived from the audit report. Nevertheless, the optimum realization of an environmental audit program (as part of an overall environmental policy) is to raise environmental consciousness of each plant to a level of daily concern, and not one strictly with a particular audit period.

Electric Utility Environmental Audits: Alberta Power

B. Peel

ENVIRONMENTAL AUDITS are one of the best ways to evaluate the environmental performance of an electric utility like Alberta Power Limited. To that end, last year the managers of our company's three major power plants were looking for some honest, measurable feedback on their plant's environmental performance.

They wanted some clear answers to basic questions: how do we stack up in terms of regulatory compliance? Emissions? Water quality? Hazardous waste management? Toxic spills? Corporate policies? How can we improve our standards? And are there any good environmental management practices that we should implement? Can they be implemented with relative ease?

Our plant managers didn't want a witch hunt and didn't want fingers pointed. They just wanted solid information, standardized measuring criteria and a chance to interpret the findings and take follow-up action to avoid future problems. They got what they wanted when, in the summer of 1989, Alberta Power became the first Canadian electric utility to use new environmental audit protocols developed by the Canadian Electrical Association (CEA), whose members include all electric utilities in Canada.

The CEA developed auditing standards and protocols in response to a growing need for consistent, unbiased audit methodology (protocols) among its member utilities. As members of the CEA, we in the electric utility industry knew that if we didn't take the lead in developing credible measures of our environmental performance—especially with the concern over global warming, air and water quality and our industry's reliance

B. Peel is Senior Engineer, Alberta Power Ltd., Edmonton.

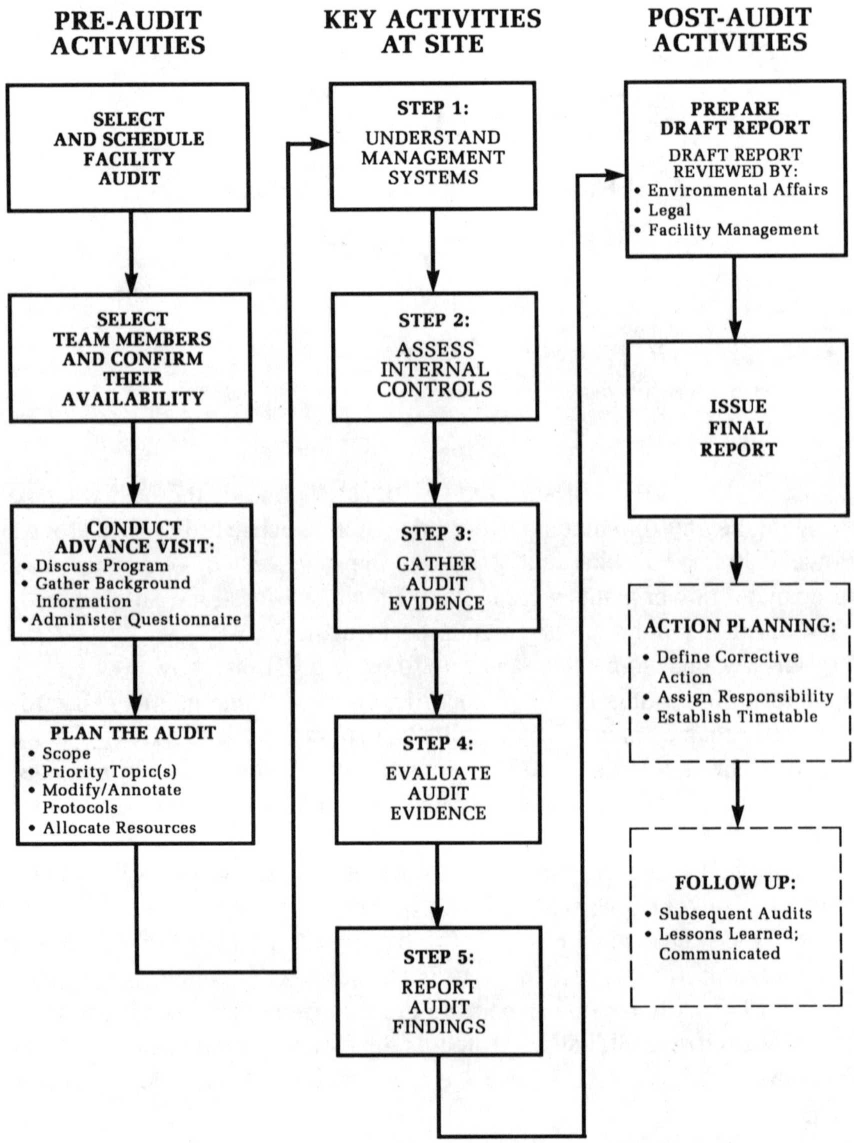

FIGURE 1. BASIC STEPS IN THE TYPICAL AUDIT PROCESS

on fossil fuels—we ran the risk of having environmental audits imposed on us by regulators and the public. But that was not the driving force behind our audits. We wanted to use environmental audits as management tools. We chose to be pro-active and responsible, and to prove that we'll help regulate our own performance.

The consulting firm of Arthur D. Little Inc. (with technical advice from NSPC and Ontario Hydro) and Alberta Power personnel helped the CEA develop protocols for measuring environmental performance; protocols we used quite successfully during our audits.

Company Background

But before we look at how we used the CEA environmental audit protocols during our actual field work, a brief background on our company: Alberta Power Limited is a sixty-three-year-old investor-owned electric utility (major shareholder is Canadian Utilities of Edmonton) with more than 1,400 employees serving in excess of 140,000 residential, farm, commercial and industrial customers throughout northern and east-central Alberta.

The company operates three major thermal generating plants, using Alberta's low-sulphur coal: the Sheerness Generating Station, co-owned with TransAlta Utilities, with two units now generating up to 764 megawatts; the older Battle River Station, with five units generating up to 741 megawatts; and the unique H.R. Milner Station, in the foothills of the Rocky Mountains, burning "junk" coal left over from a major coal mining company on the same site, and generating up to 150 megawatts of electricity. Alberta Power also operates a number of smaller gas turbine plants and remote diesel units, the latter serving small communities throughout northern Alberta.

Previous Audits

It should be pointed out that the 1989 plant audits were not Alberta Power's first foray into this area. The company has been conducting some form of internal environmental audits since 1976, the most notable being our 1985 audit of our management of PCBs. Conducted by Stone and Webster Canada Limited, the audit covered the policies, procedures and activities relating to the handling, storage and disposal of oil and equipment containing polychlorinated biphenyl in Alberta Power generation facilities and transmission and distribution equipment.

The final report and recommendations formed the basis of follow-up action that has kept Alberta Power largely out of trouble on the PCB front, even during the renewed paranoia over PCBs after the St. Basil-Le-Grande fire of 1988.

So we examined PCB management in detail, but the audits performed at our three major generating plants in 1989 were the first comprehensive

ACTIVITIES	TIME BEFORE AUDIT
Schedule the Audit	
Schedule site visit	8–12 weeks
Notify facility manager	6–8 weeks
Select team members	6–8 weeks
Assemble Background Information	
Identify information requirements	6–8 weeks
Conduct a pre-audit visit	2–4 weeks
Review background information	1–4 weeks
Develop Audit Plan	
Prepare audit protocol	2–4 weeks
Select priority topics	2 weeks
Estimate time by protocol step/topic	2 weeks
Allocate team resources	1–2 weeks
Finalize Administrative Details	
Confirm arrangements with facility	1–2 weeks
Select time and place for team members to meet	1–2 weeks
Prepare appropriate items to take to facility	1–2 weeks
Pre-audit team meeting	2 days

FIGURE 2. KEY PRE-AUDIT ACTIVITIES

environmental audits conducted at Alberta Power by an external agency. We used auditors from A.D. Little as we didn't want to have to defend the credibility of the audits to regulators or the public, should the need arise to make the audits public.

We also looked for a benchmark to conduct future audits to see where we've corrected areas that were not in compliance and what still needs to be done. We also wanted to include areas not required by regulation, where we could improve our environmental performance to demonstrate our concern for the environment.

Six Basic Elements of all Environmental Audits

According to the CEA Environmental Audit Manual for Canadian Electric Utilities, effective audit programs typically include six basic elements:

1. *Objectives*: The objectives must be specified, defined, realistic, consistent with management's needs and expectations and communicated throughout the organization.

   The objectives of our three plant audits were the same in each case, and were based on the CEA manual:

   - Compliance status of each plant with respect to applicable federal, provincial and municipal environmental laws and regulations;
   - Each plant's conformance with company and facility environmental policies and procedures;
   - The environmental management control systems which currently exist at each plant.

2. *Organization*: Effective audit programs must have strong support from upper management who are critical to achieving the objectives and obtaining accurate and reliable information. We had the complete support of upper management—in fact, the final order for the audits was made by the vice-president in charge of the plants. We were also given free rein as to how to conduct the audits.

   Choices must also be made as to type of audit and staff used. As the CEA manual warns:

   > From an organization perspective, audit effectiveness is also enhanced by the objectivity and impartiality of the auditors conducting reviews and often the independence of the audit program and operating units. If the audit program is not sufficiently independent or if the auditor's objectivity is influenced by personal beliefs or attitudes, audit results may be biased and may not provide management with a realistic assessment of environmental performance.

   We therefore chose to have the audits managed and conducted by our environmental staff outside of the operations group.

3. *Resources*: The CEA manual also warns that "the effectiveness of the audit program is a direct result of the confidence, training, expertise and proficiency of the personnel who conduct the audits." Considering both warnings, we chose to contract with A.D. Little to provide four professionals for the first audit and two for the other two audits.

   We provided observers for the first audit and participated in the second two audits. Four distinct disciplines were covered: air, water, spills, and hazardous wastes and PCBs. Prior to our first audit, I also took a course on conducting audits.

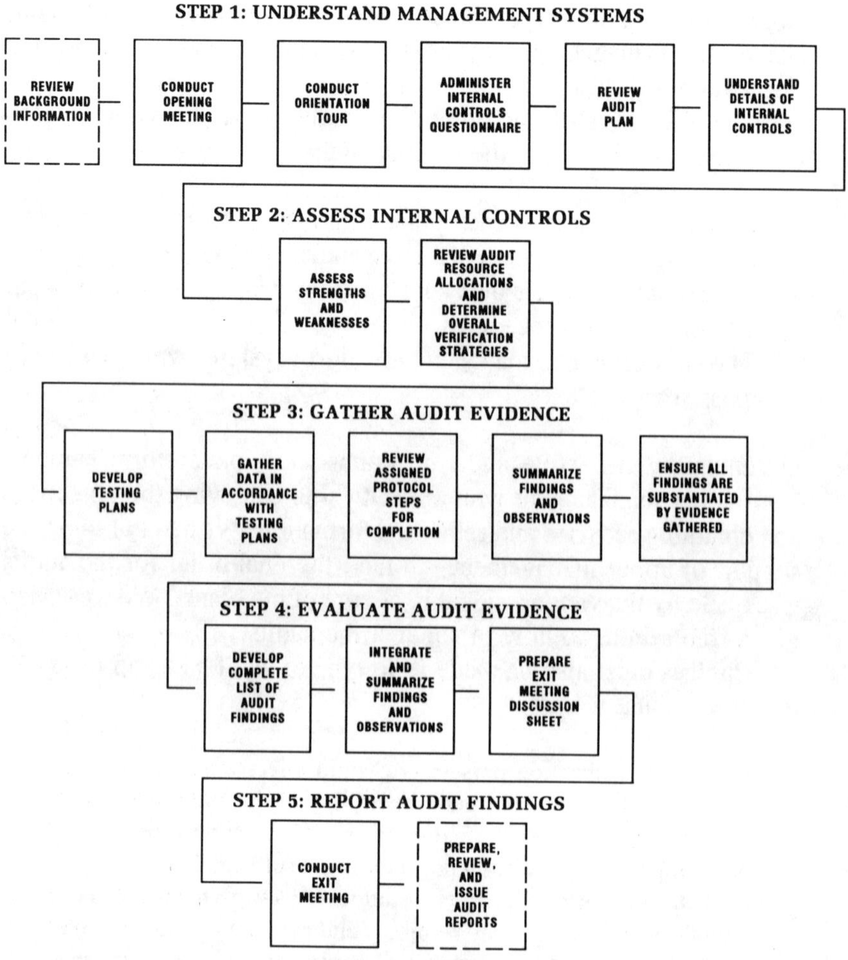

FIGURE 3. KEY AUDIT ACTIVITIES

4. *Scope*: Effective audit programs must have clearly defined boundaries for the overall scope of the program as well as for an individual audit, and these boundaries must be achievable given resource constraints. Scope includes organizational terms, functional terms, facility life-cycle terms, locational terms and risk management terms.

We chose to conduct comprehensive environmental audits of our three major plants as these pose the greatest potential threat to the environment. Besides all the facilities within our plant boundary, we included any facilities needed for the operation of the plant. For example, at our Sheerness plant, the intake pumphose (which provides water for our cooling pond) was audited, even though it is forty kilometres from the plant site.

5. *Coverage*: For companies with many sites to choose from, choices must be made as to which site or issue is to be audited, and in which order. We decided on our plants as described above.
6. *Approach*: This includes matching the nature and intensity of the audit with the program's objectives. It also includes the development of explicit instructions or audit procedures and protocols. As well, audit programs must have reporting procedures with accurate, informative written reports, sufficient documentation, confirmation that procedures were accomplished appropriately and a method to ensure timely corrective actions are implemented, as well as a mechanism to audit quality.

In a recent presentation to the Engineering and Operating Division of the CEA, A.D. Little also points out two more important considerations regarding environmental audits. First, "Audit effectiveness is ... enhanced by the development of explicit instructions or audit procedures prescribing both how the audit is to be conducted and how its quality is to be controlled." Further, Little concluded:

> Leading audit programs have, as part of the overall audit approach, an established reporting process that: provides accurate, informative written reports of audit findings; includes sufficient documentation to substantiate overall audit results; confirms that all specified audit procedures were accomplished in an appropriate manner; ensures that timely corrective actions are implemented to address identified concerns; and provides a mechanism to review audit quality via review of audit documents.

That's the overall framework for effective audit programs which we both helped to develop, and which we followed in coordinating our own plant audits. Following is a discussion of our approach, again based on the CEA manual, with more detailed discussion on key activities at the site as well as our post-audit reporting and follow-up procedures.

Pre-audit Activities

You've heard or read about pre-audit activities from previous presenters, but let me briefly outline ours:

- We selected and scheduled the facility audits for the spring and summer, roughly six weeks apart. This gave us sufficient time to complete the audit report and make any changes to our field program to improve the next audit.
- We selected team members and confirmed their availability.
- The audit manual next recommended conducting advance visits to discuss the program, gather background information and administer a questionnaire. We were able to avoid advance visits to the plants because of our familiarity with the sites. Where a site is large, however, and the audit team is totally unfamiliar with the operation, an advance visit is advised.
- Next step is to plan the audit: decide on scope and priority topic(s), modify/annotate protocols, allocate resources. Small details such as where to eat, having a room available at the site, telephones, computers and the like all had to be part of the pre-audit activities.

In addition, we examined all pertinent regulations and licenses before conducting each audit. This was done at our Edmonton office by Alberta Power audit team members. All pertinent information was sent to A.D. Little for their review prior to the field audit.

Field Work and Data Collection

The scope of each of our three plant audits was the same and included those plant activities related to air pollution control, water pollution control, spill prevention and control, solid and hazardous waste management and PCB management. Our audits did not include workplace hazardous materials or a review of radioactive sources.

The audit team in each case included Arthur D. Little auditors and Alberta Power Limited auditors, led by myself. Each audit took roughly one week, and the period under review in each case was from 1 January 1988 through the final day of each plant's audit in 1989.

Each audit review period should be a least a year, to cover the period since the previous audit. This avoids having to deal with facilities' looking good by using quick fixes prior to an audit, fearing a witch hunt. Fortunately, however, we didn't have to deal with this situation.

Twenty-five percent of our time was budgeted to be spent in preparation, forty percent on the field work, thirty percent on the post-audit report work and five percent was travel time.

Each audit was based on:

- A physical survey of the facility.
- Examination of a sample of environmental, administrative, technical and operating records available at the facility, including any submissions to the regulatory agencies.
- Interviews and discussions with key facility management and staff. These interviews were very informal. We did not record any of the discussion with video or audio equipment and no pictures were taken at the site. Only audit team members had access to our field audit notes so everyone felt freer to express their concerns without fear of retribution. This was strongly encouraged by plant management.
- Verification procedures designed to examine the facility's adherence to environmental laws and regulations and company policies.

All deficiencies noted during the audit were discussed fully with representatives of each plant and Alberta Power management. As the final report for each audit noted, we followed the audit protocols established by the CEA.

As well, our reports warned that "efforts were directed towards sampling all facets of environmental performance in the period covered, but it is important to recognize that such a sampling method can only support general conclusions and does not necessarily identify all potential problems."

Key activities on site during data collection included four steps: understand management systems, assess internal controls, gather audit evidence, evaluate audit evidence and report audit findings.

To understand management systems, we held opening meetings and orientation tours, as well as administering an internal controls questionnaire. We then interviewed key staff, conducted additional tours with them and reviewed practices and procedures.

In step two, we assessed the strengths and weaknesses of the design of each facility's programs, practices and control systems and developed verification strategies for those systems. For example, a particular plant may not have formal training to deal with areas where non-compliance clearly reflects lack of knowledge, perhaps about instrument calibration.

Further, we checked whether personnel were trained and experienced and had clearly defined responsibilities. We also looked at the system for project review and approval and internal verification procedures. And we assessed protective measures, emergency response plans, alarms, spill response, environmental controls and practices, and record keeping systems—especially for their accessibility, comprehensiveness and compliance with provincial regulations and company policies.

To gather audit evidence, we interviewed key staff again, reviewed documents, tested systems and procedures and observed practices. When evaluating audit evidence, we first of all confirmed that sufficient evidence had been obtained, then summarized audit expectations and observations.

In reporting audit findings, we shared these with plant management at the end of each day. Serious things were to be dealt with each day—fortunately, we had only one such minor finding requiring instant action, which we reported to plant management. They proceeded to have it corrected immediately. Findings were written up at the end of the week in draft form and shared with plant management. We went through each finding and our suggested recommendation in our draft report to get comments and suggestions and clarification where necessary, and then went away to write our final report.

A host of recommendations were made and acted upon. For example, at our newest plant, Sheerness, the audit identified the need for improvements in monitoring equipment maintenance, reporting procedures, on-site waste management and hazardous goods spill response. Most changes should be finished by the end of 1990.

Following is a sample item and recommendation from the final audit report for Sheerness plant:

1.5 *Equipment Calibration*

In reviewing the calibration of air pollution monitoring systems, we noted the following:

a. The high volume samplers at both the pollution trailer and the cooling pond valve house have not been calibrated once every three months, as recommended by Alberta Environment. Draft AMD 86-1 Sections 1-d (vii and viii).

b. The wind and speed direction system has not been calibrated in the past four to five years. Alberta Environment has suggested that the wind speed and direction systems be removed for mechanical inspection, electronic alignment and calibrated at least once every two years. Draft AMD-86-1 Section C.1.1.

Suggested Recommendation: Review the suggested method of calibration of equipment as outlined in the Draft AMD-86-1 and develop a written procedure that includes a regular schedule for all monitoring equipment with assigned responsibilities for calibration and retention of calibration records.

Following is the resultant action, as detailed in the Sheerness plant Environmental Action Plan (the original also indicates who is responsible and what the deadline is):

1.5 *Equipment Calibration*

- (*i*) Procure calibration equipment for high volume samplers and set-up standard job request for routine calibration (every three months).
- (*ii*) Set-up a standard job request and perform routine calibration of windspeed and direction equipment (every two years).
- (*iii*) Set-up an environment calibration records file.
- (*iv*) Record calibration dates on all related Alberta Environment reports.

Post-audit Activities

According to our CEA manual, post-audit activities typically should include:

- preparation of draft report; reviewed by environmental affairs, legal, facility management;
- issue final report;
- action planning: define corrective action, assign responsibility and establish timetable;
- follow-up: subsequent audits, lessons learned and communicated.

For Alberta Power, following each audit and the sharing of draft reports, the audit team produced a final report of between eighteen and twenty-two pages, concisely listing all deficiencies noted, as evidenced in the sample above from the Sheerness plant final audit report. Suggested recommendations for corrective action were also noted in each case. This was done within two weeks of receiving comments from the plants and no more than four weeks following the field audits.

Since the plant people know best how their plants operate, the management of each plant puts together their own specific action plans to address the deficiencies. Though some admitted they were leery of the auditors at first, they found them to be professional and objective. As one manager reported, "In the end there weren't too many surprises in the report findings."

Subsequent action plans for our plants range from minor record-keeping changes to capital works projects of up to $500,000. As well, some parts of some licenses need to be revised or amended.

One plant manager found the most overwhelming task to be organizing all information contained in the audit. He and his management team found that "one small recommendation turned into literally hundreds of mini action tasks, so we decided to organize the whole plant on a computer. Now we have a total station action log, detailing all action plans for the next four years." They're also looking at waste recycling, "green purchasing," biodegradable waste—even alternatives to styrofoam cups.

Employees contributed openly during the audits and are to be commended for their candid comments. In fact, employee honesty is probably the reason Alberta Power enjoys such a good relationship with regulators. The federal and provincial government agencies consider us quite proactive in not waiting for regulations to be imposed.

Recommendations Based on our Experience

There are a number of changes we'll make next time we do environmental audits of our plants—in fact, the next audit is already in the planning stages:

1. More pre-audit preparation is needed by all team members.
2. A follow-up audit to check action taken on recommendations should be conducted.
3. Records should be more accessible. They were hard to get our hands on as different people had them.
4. Scheduling time with key people at the plants was difficult, with holidays especially. With more notice given—and knowing who to talk to now—this should be overcome during our next audit.
5. We're going to take along a computer and write up results the same day on each day of any audit.
6. A greater commitment to finish the final report will be needed.
7. We're also going to ensure the plants commit to sharing their action plans with us earlier on than they did this time.
8. Staff time and staff commitment to the project was problematic at times. Also, the proper level of staff is something we'll look at more

closely, especially when it comes to writing the report. The final report is one of the most important parts of the audit, as the action is based on it. We intend to use external consultants on all plant audits for the foreseeable future as we don't have sufficient qualified staff with the time to devote to such endeavours.

Conclusion

Environmental audits are now a positive, proactive and effective management tool at Alberta Power Limited. They were well-received by our plants and appropriate follow-up action has been taken in most cases for most recommendations. Another round of audits is being planned. In fact, environmental audits at our plants will now occur at the call of plant managers. They may even require more frequent audits than our senior management first anticipated. Done properly, they've become another valuable management tool.

Reference

A.D. Little Inc. (1989). *CEA Environmental Audit Protocols.* Edmonton, Alberta.

Environmental Audit of Industries in the Junction Triangle

B. Singh

THE JUNCTION TRIANGLE is an industrial/residential area of approximately 6,000 residents in the northwest section of the City of Toronto. It was settled a century ago and derives its name from railway "junctions" and the confining triangular boundaries formed by the railway tracks.

Early land development practice of locating homes next to factories, i.e., residence next to workplace; changes in production and industrial materials; mobility of the work force; changes in transportation from rail to road; and society's demand for a cleaner and healthier environment provided ideal conditions for environmental conflicts in early 1982. The Junction Triangle environmental audit in 1985 was undertaken to address and resolve these conflicts. It evaluated the environmental quality of the neighbourhood and the perceptions of the citizens, and provided assurances about the environmental management practices of the companies and regulatory agencies.

Environmental Problems

In the early hours of 6 April 1982 fumes from the sewers in the Junction Triangle entered homes and sent seven persons to hospital. The area was cornered off and neighbourhood schools were closed. It took much effort and ten days to find the cause of the problem. On 16 April, an area industry announced that it was responsible. An employee, in order to cover-up an operating error, had discharged approximately 800 gallons

B. Singh is an Environmental Relations Consultant with Environmental Relations Consultant Inc., North York, Ontario.

of a mixture of very odorous material to the sewer system. The company was charged for violation of the Environmental Protection Act, and was found guilty and fined in November 1982.

The discharge confirmed in the minds of several citizens that dumping of chemicals into the sewers was the prevailing practice of area industries, sanctioned by management as a cheap way of "treating" wastes, and perhaps even condoned by regulating agencies through lack of stringent policies of industry activities. The ten-day delay between the incident and the discovery of the cause generated much media and political pressure about reprehensible practices of industry and the lack of control by government. Immediate action to avert serious disaster in the neighbourhood was called for during the period of investigation.

Abatement Action and Public Reaction

In response to calls for stringent action to prevent similar incidents, the Ministry of the Environment took the lead role in co-ordinating action plans of all regulatory agencies for the following area industries: Glidden, Nacan, Inmount, Anchor Cap and Closure, Viceroy and Canadian General Electric. The message was clear: clean up or close the plant. These action plans resulted from "environmental audits" of a less formal character than what is being advocated at this symposium.

The audits resulted in voluntary pollution abatement programs at all of the companies. Nacan, Inmount and Glidden were required to prepare more extensive plans of action. The other industries were deemed to have completed all works identified by the informal audits and to be in full compliance with regulations.

Nacan and Inmount submitted programs to the Director to control discharges in accordance with Section 9 of the Environmental Protection Act (EPA). Glidden considered that it was in compliance with all regulations and requirements, a position with which the ministry disagreed. The ministry carried out an "environmental audit" by having a Provincial Officer undertake an inspection under Section 126 of the EPA and prepare a Provincial Officer's Report for the purpose of serving a Control Order under Section 116 of the EPA.

In February 1983 the ministry issued a Control Order to Glidden and Program Approvals to Nacan and Inmount to undertake works that were more protracted and significant. All works were carried out in accordance with these documents.

In spite of intensive activities during the spring and summer of 1982 and the years immediately following, however, the area continued to be plagued by a number of odour incidents from emissions to the air or smells from the sewers. These incidents generated many public meetings and the formation of a second citizen group called the Watchdog Committee. There was already an existing group called the Junction Anti-Pollution Group. Rivalry between the two groups, with membership from two rival political parties, did not make matters any easier.

Environmental Liaison Committee

The public meeting on 24 June 1982 was charged with great emotion and took the form of industry and government bashing with little constructive ideas other than to shut the industries down. It was obvious that a calmer forum was required for understanding problems and formulating solutions. I suggested the formation of a liaison committee of citizens, industries and governments to meet at regular intervals to review the environmental situation and develop appropriate abatement and improvement plans.

Shortly thereafter, I was invited to speak to seven industry representatives about the workings of the proposed liaison committee. Several of the industrial representatives were sceptical about being able to establish useful dialogue with the citizens or their political representatives. Their experiences with the public had been confrontational and stressful; however, I was able to convince them that the prevailing climate of confrontation and abuse had to change to one of cooperation and harmony if they wished to continue to operate in the community. It was agreed that anything would be better than the current situation and that a means of establishing the committee should be found.

With support from the City of Toronto Planning Department and the Ministry of the Environment (MOE), the Junction Triangle Environmental Liaison Committee came into being in July 1983 with myself as the first chairman. As envisaged, the committee provided a useful forum for the exchange of ideas and the review of progress in pollution abatement. We were gradually able to replace *confrontation* with *cooperation,* and *fear* and *suspicion* with *trust* and *respect.*

Meetings were and are still being held every two months. An interesting development at the inception was the decision of the two citizen groups that they did not wish any elected persons as members of the committee. The citizens wanted to deal directly with the companies and government officials. They wanted issues kept in perspective and did not wish to have the committee used to further political careers of individuals. If political action became necessary, the citizens would seek political assistance as needed.

Abatement activities continued at a heightened pace but problems, especially odour complaints and concerns about health, continued to plague the area. Each incident attracted media attention with banner headlines and full radio and television coverage.

Certain citizens have stated that they would never be satisfied until the industries are relocated. To preserve employment and its industrial tax base, the City of Toronto would like to maintain industries in this neighbourhood. Neighbourhood conflicts are likely to continue for some time and the Liaison Committee is assured of a continuing role as industries are replaced on the currently zoned industrial lands.

Perception and Reality

Differences in perceptions developed in the liaison committee between the citizens on the one hand and the industry and government members on the other. The citizens stated that enough had not been done and that significant improvements were still needed. Industry and government members stated that they had done all that was practicable under the circumstances. In the spring of 1984, the Minister of the Environment toured the area to see conditions for himself. During the tour the citizens made it clear to the Minister that in spite of all that had been done, they did not consider that the pollution problems had been resolved.

The Minister instructed that something had to be done about the perception of the citizens. He advanced the thesis that "Perception was reality" and that the ministry was required to deal with this "reality." I was faced with accepting the citizens' "perception" as "reality" and formulating action to deal with this "perceived reality" or to change their perception. The challenge was to establish a "reality" that was acceptable to and accepted by all parties. I was convinced that no one would benefit from actions based on "false realities." A way had to be found to arrive at a common "reality" upon which to base future abatement action.

Independent Environmental Audit

As my one-year term of chairmanship was ending in July 1984, I used the occasion to review the committee's achievements and to allude to the gap that existed between the perceptions of the citizens and the other members of the committee. I suggested that an "environmental audit" by an independent consultant to examine the situation would resolve the difference in perception. I had discussed environmental audits with John Reed, presenter of another paper at this symposium and decided that the term "environmental audit" would appeal to the committee as a new plan of action. Use of an independent consultant and full public disclosure were new elements in the suggestion for the environmental audit.

This suggestion, like most new ideas, was greeted with customary coolness. As the meeting progressed differences in perception about pollution abatement actions and their effectiveness arose. It did not take long for members to start voicing support for the idea of an independent audit as a means of resolving the differences of perception. I had to make it clear that my idea was not discussed with my superiors, and was only put forward for discussion and consideration by the committee.

At the next meeting, the committee passed a resolution asking the Minister of the Environment to fund an environmental audit, by an independent consultant, of all the Triangle industries with known or suspected problems. I was requested to continue in the chair for another six months to develop and lead the committee through the audit. The industry members supported the motion subject to the condition that all identified industries agree to be audited. All of the industries agreed and the Ministry of the Environment funded the audit.

Scope of Environmental Audit

The audit was a completely open exercise with equal participation by the citizen, industry and government members of the liaison committee. A subcommittee with balanced representation from the citizens, industries and government was struck to establish the terms of reference and to select the consultant. Prior consensus was considered necessary for the success of the audit and this was achieved. The first six pages of the terms of reference for the environmental audit are given in the Appendix.

The auditor was required to look at four factors and their interaction with each other:

- the quality of the environment
- the performance and impact of the industries
- the performance of the Ministry of the Environment
- the expectations of the citizens.

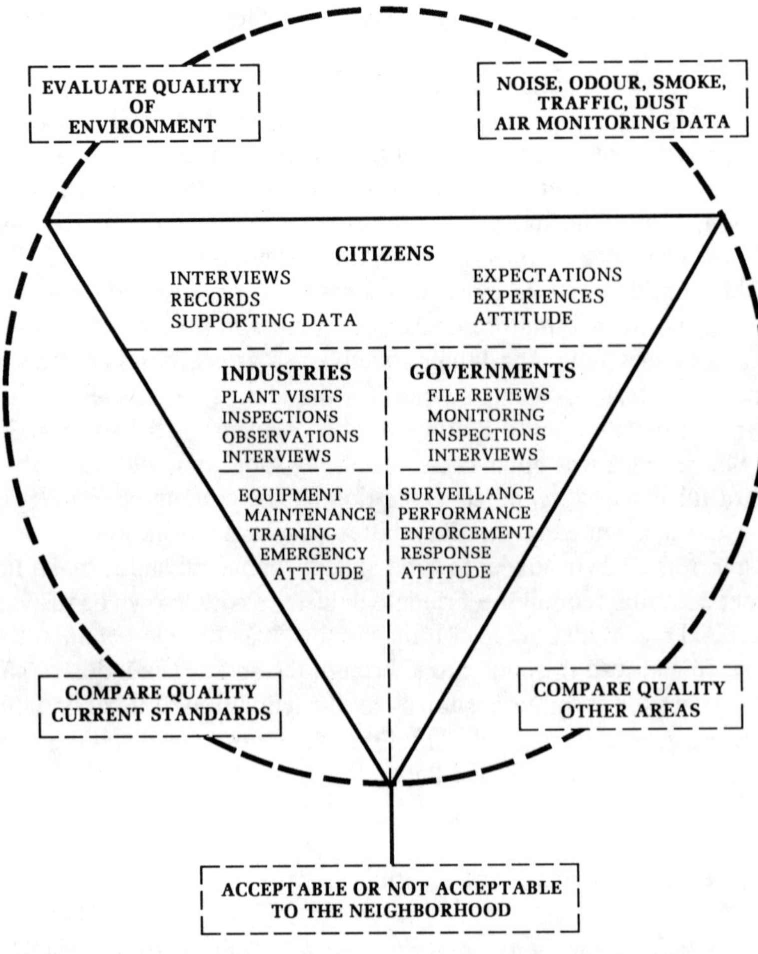

FIGURE 1. JUNCTION TRIANGLE ENVIRONMENTAL AUDIT

Citizens, industries and government interact with each other and must act together to establish what is an *achievable* and *acceptable* quality of environment for the neighbourhood.

Citizens must be able to live in a hazard-free environment and are entitled to the enjoyment of life while residing in the neighbourhood. They were to be interviewed about their concerns, perceptions and expectations about their environment. The industries produce goods which are sold at a profit; they generate direct and indirect employment; they pay taxes and they have an impact on the environment with noise, odour and traffic, as well as visually and aesthetically. The government must ensure that all legislative requirements are observed and that the well-being of the citizens and the environment is protected.

The Junction Triangle Environmental Audit contained several unique features:

- It was the first time that citizen groups had a role in setting the terms of reference for a study of this nature and were involved in the selection of the consultant and review of the draft report prior to finalization.
- The consultant did not only look at the industries, but at the quality of the environment and compared it with other comparable areas in the City of Toronto.
- The ministry and the federal monitoring programs were examined for their adequacy.
- The performance of the ministry in meeting its regulatory responsibilities was another feature of the study.

A sampling of citizens' perceptions and observations was an integral part of the study. The citizens were divided into two separate groups, those who were active on the committee or associations and the citizens who were not. These were considered separate sample groups and generated differing comments and concerns.

Audit Findings and Recommendations

Sample quotations from the Conclusions and Recommendations in the Executive Summary of the Junction Triangle Environmental Audit report are given below to illustrate the environmental situation at the time of the audit which was carried out in the latter part of 1985.

> CONCLUSIONS:
>
> To become aware of the residents' concerns and opinions, random samples of residents living in the Junction Triangle and from another, similar area of the City were asked a series of questions about the environmental quality of their neighbourhoods. The results indicate that environmental quality is viewed as more of a problem by Junction Triangle residents. Dust, odours, and smoke are the most frequently reported concerns. Factories and traffic are perceived as the main causes of the problems.

Most Junction Triangle residents support the pollution abatement efforts of the MOE and more than three-quarters of the residents think that the MOE complaint response system is adequate.

The air quality of the Junction Triangle is not much different than that measured at other stations in Metropolitan Toronto....

At the twelve industries evaluated, liquid and solid wastes are properly managed, response plans have been developed for accidental releases, and worker training programs for safety and proper handling of chemicals have become more comprehensive.

Based on predictive calculations, eight of the industries have controlled emissions that may cause odour incidents while complying with point-of-impingement concentrations....

The MOE has played an important role in bringing about improvements to the environmental quality of the Junction Triangle despite difficulties with translating the Environmental Protection Act and Regulations into abatement activities.

While it was once suggested that the MOE should establish an area office and staff it on a 24-hour basis, such an office would be an inefficient use of limited resources.

RECOMMENDATIONS

Educating managers and workers about the involvement in the reduction of fugitive emissions.

Industries should continue their research and development efforts with respect to replacement of some currently used raw materials with compatible less-odorous materials.

Develop a realistic objective for odours in the Junction Triangle recognizing that the total elimination of odours is not a practical objective.

Complainants should always be informed of the results of investigations.

Use of Public Environmental Audit

The public environmental audit used in the Junction Triangle has application in those situations where the quality of the environment of a neighbourhood is affected by the collective impact of a concentration of industries and requires the concerted efforts of all industries to bring about significant and appreciable change.

A public environmental audit goes beyond the narrow responsibility of an environmental audit for a single company. It is not only concerned with the company complying with regulations and having good operating and emergency response programs in place; it evaluates the environment of the neighbourhood and sites to establish with the citizens an acceptable quality for the environment of that neighbourhood. It fosters and promotes cooperation among competing industries who share the same neighbourhood and collectively impact on its quality.

The exercise demonstrated that it is possible for citizens, industries and government to work together to examine and understand environmental issues and develop plans of action for the resolution of problems.

References

Barney Singh. (June 1983). *Industrial Abatement in the Junction Triangle.* Industrial Waste Conference, Toronto.

Barney Singh. (June 1985). *Update on Toronto's Junction Triangle.* Industrial Waste Conference, Toronto.

Robert E. Millward. (January 1990). *Bloor-Junction Area Land Use Policy Review.* Toronto.

Appendix: Request for Proposal for Junction Triangle Environmental Audit

Terms of Reference

I. THE STUDY

An Environmental Audit in the Junction Triangle area of the City of Toronto will consist of:

- An evaluation of the abatement management programs of identified industrial plants.
- An evaluation of the impact of area industries on the quality of the environment in the neighbourhood.
- An evaluation of the Ministry of the Environment's abatement management programs in the Junction Triangle.
- An evaluation of the Ministry of the Environment's complaints response system.
- An evaluation of the environmental quality of the neighbourhood by evaluation of air monitoring and plant production data.
- An evaluation of citizens' expectations as it relates to quality of life when compared with current accepted standards.
- A report of all evaluations carried out.

This study will not delve into health as the City Department of Health will be undertaking a concurrent health study of children (Health Assessment Program). Citizens, bringing health concerns to the attention of the consultant, should be directed to the Western Area Health Unit of the Department of Health of the City of Toronto.

II. DESCRIPTION OF WORK TO BE DONE

A. *Industrial Operations*

Detailed surveys of twelve plants will be carried by the consultant; however, if other operations within or outside the Triangle are considered to have an impact on the quality of life and the environment, the consultant should identify these operations and impacts, and make recommendations, as he sees fit, for further evaluation of these situations. Odours from the sewer system should be addressed in the study.

The twelve (12) plants are:

1.	American Standard 1401 Dupont Street	ceramic and bath fixture manufacturing
2.	American Standard 80 Ward Street	fibreglass bathtub manufacturing
3.	Anchor Cap and Closures 275 Wallace Avenue	bottle caps manufacturing
4.	Canada Packers 60 Paton Road	gelatine manufacturing
5.	Canadian General Electric 940 Landsdowne Avenue	resin manufacturing
6.	Glidden Company 351 Wallace Avenue	paint manufacturing
7.	Inmont Company Inc. 303 Campbell Avenue	ink and coatings manufacturing
8.	ITT Grinnell 2440 Dundas Street West	iron foundry
9.	Nacan Products Ltd. 371 Wallace Avenue	adhesives and resin manufacturing
10.	National Rubber 394 Symington Avenue	rubber floor mats manufacturing
11.	Royce Dupont Poultry 1526 Dupont Street	chicken packing
12.	Viceroy Rubber 1655 Dupont Street	rubber goods manufacturing

The consultant will develop an understanding of the operation of each company and its impact on the environment by discussion with Ministry and company staff and through plant visits. Plant visits may be pre-arranged with management or be impromptu and may be accompanied by Ministry staff as required. If required by the company, discussions and interviews of employees may be undertaken only in the presence of management.

The consultant shall do the following:

1. Hold discussions with Ministry staff and review Ministry files for past and current abatement activities on each plant.
2. Visit plants and,

   - identify each process stream and its potential for contaminating the environment;
   - identify contaminant and waste emission sources within the plant;
   - review all control equipment for suitability and effectiveness;
   - review the maintenance and operation of all control equipment to ensure proper functioning;
   - review operator competence for the operation of all emission control equipment;
   - review materials handling and storage including control of vapours during the filling operations of volatile substances;
   - assess all spills and contingency measures (manuals, staff readiness, ready availability of materials and equipment);
   - assess all storage and handling of hazardous and dangerous materials;
   - identify waste generated, waste handling practices for conformity with legislative requirements for the handling of wastes.
   - interview company representatives to obtain their perspectives.

The assessment of the company's operation must be carried out against legal requirements, recognized good practices and technical achievability and must contain comment on management attitude towards pollution control.

B. *Ministry Operation*

The consultant must develop an understanding of the Environmental Protection Act and the regulations that apply to the control of industrial enterprises in order to understand the scope and role of the Ministry's abatement program.

The consultant shall do the following:

1. Assess Ministry performance through a file review of abatement programs for various plants with particular emphasis since April

1982, i.e., review whether enforcement of abatement has been consistent with legislative, requirements, good practices, recognized standards and technical achievability.
2. Identify all outstanding abatement requirements and status, i.e., whether being currently pursued or have not been identified or identified and action being developed.
3. Review the Junction Triangle Alert Plan and its effectiveness. This may involve reviewing the complaint response of the Fire Department and the Works Department of the City of Toronto and Metropolitan Toronto.
4. Review the Ministry's complaint response system and Duty Officer's system for dealing with complaints; and identify required changes.

The Ministry's files on abatement activities and complaints will be made available to the consultant and Ministry staff and will be available for discussions of any past and current abatement activities.

C. *Evaluation of the Quality of the Environment*

The consultant will evaluate the quality of the environment by review of air quality monitoring data and by review of the perspectives of citizens through review of complaints and an evaluation of complaints.

The consultant shall do the following:

1. Review data for standard pollutants for air monitoring station at the Perth Avenue School and compare with two other Metropolitan Toronto stations (all data are on computer files and will be made available to the consultant in analyzed format). The consultant is not expected to collect data or to carry out computations but only to do an assessment of quality by comparison with other stations and against recognized standards.
2. Assess impact on the quality of the environment by reviewing the Ministry complaint records to determine statistical patterns of complaint, i.e., geographical, by complainant, type of complaint, time of complaint, etc.
3. Assess industrial impact by reviewing real time monitoring data, production data and meteorological data that will be generated during a ten to fifteen day period. The Ministry will operate a

mobile monitoring unit during this period and prepare a technical memorandum of the data for the consultant's review. The technical memorandum will be an appendix of the completed report.

4. Review data from joint Federal Provincial station at Perth Avenue School for volatile organic compounds and assess the significance.

D. *Citizen Expectations and an "Acceptable Quality of the Environment"*

A review of abatement measures and measurement of contaminants in the air cannot alone provide a complete picture of the quality of the environment. The concerns and expectations of the citizens living in the area must be examined as a component in the admittedly subjective decision as to whether an "acceptable quality of the environment" has been achieved.

It is expected that the consultant will develop an understanding of these concerns and expectations through interviews with citizens and citizen groups, long term and newly arrived residents and prepare a review of material addressing their concerns.

The consultant shall do the following:

1. interview a "representative sample" of citizens not actively involved in citizen groups to obtain information and perspective;
2. interview the executives of the Watchdog Committee and the Junction Triangle Anti-Pollution Group to obtain their perspectives;
3. assess the citizens' request for a 24-hour Ministry presence in the Junction Triangle to respond to citizens complaints and to provide better surveillance of industrial activities.

The society we live in is much different from the one in which the community first developed. Dramatic changes in chemical technology have created hazards unknown 100 years ago, while an increasing awareness of environmental concerns has changed the attitude of citizens towards their industrial neighbours.

With all of the foregoing information in hand, the consultant should be able to provide an evaluation of the present situation and make recommendations for improvement, thereby lessening the conflicts that have persisted in this community for over a decade.

III. PROPOSAL CONTENTS

The proposal must include:

1. The consultant's interpretation of the requirements of the study and the approach that will be employed in meeting the Terms of Reference.
2. A schedule of tasks and estimated time to carry out each task.
3. A description of team structure and accountabilities of team members.

Environmental Auditing: A Small Company's Perspective

J.M. Shepherd

NACAN PRODUCTS LIMITED is a manufacturer of specialty chemicals. We make adhesives, synthetic polymers and starches in six plants across Canada. Some of these are relatively small plants, with an average size of about forty-five employees. In our company, the primary responsibility for safety and environmental affairs rests with operating management. I provide staff assistance, as Divisional Vice-President of Engineering and Management Services.

Nacan is a wholly owned subsidiary of National Starch and Chemical Corporation, a company ten times our size, headquartered in New Jersey. Our parent company has thirty-five plants in the U.S., most of them small. It has a small Safety and Environmental Affairs Department, and takes a similar approach to managing safety and environmental matters.

Since 1987 we have been conducting safety and environmental audits. We prefer to call them "reviews" (although I will use the word "audit" today) because to us, the term "audit" implies a formal comparison, against generally accepted principles, with some degree of precision and authenticity. We do not feel that what we do should be called an "audit"; our program is still evolving. Although we do combined audits for health, safety, and the environment, I will focus here on the latter.

Why a Small Company Should Do Audits

Better Compliance. As managers, we need continual assurance that we are in legal compliance, not only in letter, but also in the spirit of the

J.M. Shepherd is Vice President of Nacan Products Ltd., Brampton, Ontario.

law. We need to know that there will be no surprises. We need to have a minimum of vulnerability.

Nacan had an out-of-compliance incident about ten years ago that could have been prevented, had we been carrying our environmental audits at that time. In hindsight, it was basically a question of a site manager's lack of awareness of legal requirements. We believe that an audit can help to focus an operating manager's attention on the legislation, and that this leads to better compliance.

Audits often point out the need for improved operating records and for improved documentation. Often, where there is an intermittent out-of-compliance situation, good documentation is essential for a proper analysis of the problem. In most companies, smaller plants often have poor documentation.

Communicate Management Endorsement of Policy of "Responsible Care." We are one of the smaller members of the Canadian Chemical Producers' Association (CCPA). As a condition of membership in that association, we are committed to the principles of responsible care in the management of chemicals. We cannot allow our operations or our products to present an unreasonable risk to the public, to the environment, or to our employees. We are committed to complying in letter and in spirit with all relevant legislation. We have adopted six extensive industry codes of practice, covering cradle-to-grave management of chemicals. We would find it extremely difficult to meet these obligations without audits, especially since our small plants lack full-time environmental specialists.

Identify Risks and Hazards That Go Beyond Questions of Compliance. One of the ways that we have found to identify environmental hazards is through the audit process. We then conduct what we call a HAZOP (hazard and operability review) study. Over the past three years we have carried out about fifteen of these studies. Most of them have led us to make significant capital expenditures. Even if we sometimes determine that little can be done to minimize a particular hazard, the HAZOP exercise often suggest improvements in emergency response procedures.

Improve Community Relationships and Public Image. Some of our plants are in small communities. One of the smaller plants is in west-central Toronto in an environmentally sensitive area close to housing. In that area we have participated for over seven years now in an environmental liaison committee, meeting with local residents' groups on a regular basis. Residents want to be assured that a company is conscious of its environmental responsibilities. In fact, the committee, in 1985, with MOE funds and guidance, evaluated proposals from four independent consultants, and chose one of them to do an environmental audit for each of the ten firms in the area, including our own plant. The results were reassuring to ourselves, as well as to the community.

Planning and Developing a Small Company Program

There are four basic steps:

1. Define goals and objectives:
   - management expectations
   - regulations to cover
2. Determine:
   - degree of written vs. oral reports
   - frequency of audits
   - degree of legal involvement
   - distribution of reports
   - audit team composition and size
3. Prepare materials:
   - environmental (and safety) standards
   - pre-visit questionnaire
   - audit check list
   - audit schedule
4. Determine follow-up procedures:
   - remedial/corrective action responsibilities.

Goals and Objectives

Our goals are as follows:

1. Independent verification of compliance:
   - with regulations
   - with corporate policies and procedures
   - with good operating practices

2. Assurance to management that:

   - operations are consistent with good practices
   - systems are in place and operating
   - legal and ethical responsibilities are being met

3. Provide information to:

   - help managers improve performance
   - give management overview of operations
   - identify potential issues/problems

4. Save money by:

   - preventing losses
   - anticipating future costs—budget planning

5. Provide means to:

   - avoid/solve/anticipate problems
   - show "due diligence."

Regulations and Practices to Cover (Environment Only)

1. Provincial Environmental Protection Act(s) and regulations, covering air emissions, waste management, spills.
2. Federal Transport of Dangerous Goods, covering labelling, documentation, training, and spill contingency planning.
3. Federal Environmental Contaminants Act reporting for new chemicals (to be replaced by Canadian Environmental Protection Act—CEPA).
4. Municipal by-laws on discharges to sewers.
5. Undertakings through Ministry of Environment Program Approvals.
6. Undertakings from permits for installation of environmental control equipment.
7. Storage of PCBs.
8. Presence of asbestos.
9. Internal procedures

   - underground storage tank monitoring
   - emergency response planning
   - spill contingency plans

- incident reports
- waste management

Procedural Matters

1. *Report Format.* After completing the audit, and before writing the report, it's important to provide an oral briefing to the local site manager.

   We believe that the report should show good performance as well as out-of-compliance situations, and provide an accurate and balanced picture of total compliance.

   Where possible, try to include recommendations to fix the stated problem, based on discussions with local operating personnel. However, it is important to document only recommendations that are known to be implementable by the facility. If an out-of-compliance situation exists without an immediate implementable recommendation, obtain legal assistance in writing the report. Corporate liability is at its worst when compliance problems are identified, and no solutions are planned.

   Be factual and quantify statements where possible. Try to avoid the use of adjectives like "very," "significant," "high," "negligent," "dangerous" and "neglectful," which can be misinterpreted.
2. *Frequency of Audits.* We audit each site once a year. Experience shows that non-compliance situations can arise quickly.
3. *Degree of Legal Involvement.* If out-of-compliance situations are identified, which do not have an immediately implementable recommendation, our parent company legal department must be advised *before the report is written.*
4. *Distribution of Reports.* Copies are sent to the company president, the local plant manager, the manufacturing vice-president, the safety and environmental affairs vice-president at our parent company, and me.
5. *Audit Team Composition.* Audit teams consist of one to two persons from another plant location, working on-site for two to four days, with full-time assistance from one local person. Possible members are operating managers or safety and environmental personnel from other plants, or people from our corporate or parent company safety and environmental affairs staff. In a smaller company such as ours, most of these people usually have other additional operating responsibilities.

Audit Materials

1. *Corporate Environment Standards.* Safety and environmental "minimum practice" standards have been prepared in twenty-six areas. Although most of these are safety related, the environmental ones include waste management, underground storage tank monitoring, emergency response procedures for spills or environmental upset, and incident reporting. As time goes on, additional standards are developed through a consensus process with plant management. An audit questionnaire is based on these standards.
2. *Legal Requirements.* Each plant site must have an up-to-date "Blue Book" file containing copies of all safety and environmental legislation pertinent to operations at that site. The Blue Book must include copies of any Ontario "Program Approvals," Ministry directives, permits for installation of environmental control equipment, with copies of any undertakings the company has agreed to carry out.
3. *Internal Procedures.* Each plant site must have an up-to-date crisis management "Red Book" containing copies of local safety and environmental emergency procedures relating to fire, spill, run-away reaction, handling of dangerous goods, flood, bomb threat, etc., along with a hazardous materials inventory.
4. *Pre-Visit Questionnaire.* A pre-visit questionnaire is sent to the audit site which:

   - Identifies personnel responsible for air emissions, waste water discharges, waste management, PCBs, unexpected adverse effects.
   - Describes the site, operating schedule, process modifications planned, describes major pollution control facilities, identifies any non-compliance during the past year.
   - Highlights major environmental concerns or areas of regulatory uncertainty.
   - Identifies what records are available or are not available, with reference to specific legislation and corporate standards.

5. *Audit Check List.* In addition to the questionnaire referring to the Corporate Safety and Environmental Standards, and a review of the recommendations from previous audits, a check list has been developed to insure that important areas are covered by the audit team. As new safety and environmental legislation is enacted, this check list will be revised (MISA, Ont. Reg. 308, CEPA, Quebec depollution

certificate, etc.), but currently includes the following major environmental topics:

- Blue Book adequacy
- Red Book adequacy
- Ministry orders, program approvals
- Ministry permits for new construction
- Process changes since the last audit
- Review previous year's appropriation requests for SEA checklist
- New raw materials introduced since the last audit
- Anticipated process changes
- MSDS files on substances handled at the site
- Review of incident reports for spills, accidents, emissions
- Were reports of incidents timely and to proper authorities?
- Are there any situations involving chemical production, use, storage or disposal that might be termed imminent hazards? Have these been subjected to HAZOP analysis?
- Emergency response capability—on-site and for goods in-transit
- List of bulk and T/L materials purchased FOB point of origin, with suppliers' emergency telephone numbers
- Status of Community Awareness and Emergency Response (CAER) program
- Continuous emission monitors, repairs, adjustments, calibration
- Control of non-point sources (loading and unloading operations, plant site run-off, in-plant transfer)
- Spill prevention control plans
- Underground storage tank monitoring for leakage
- Operation and maintenance of sewage pretreatment facilities
- Sewage effluent permits and limits, with record keeping
- Waste recycling programs
- Waste management generator and handling permits
- Waste disposal contracts
- Waste handling, record keeping, including age of waste at time of disposal. Inspect storage areas
- Review of solid waste disposal practices
- Inspect storage area for PCBS.

6. *Schedule.* The audit is scheduled months ahead of time, to insure that personnel responsible for the various areas are on site, and have the time available to work with the auditors.

Follow-Up Procedures

Audit findings are first reviewed verbally, during a debriefing with the local plant manager, before the audit team leaves the site. Each recommendation is assigned to a specific person, who has responsibility for its completion by a specified date.

A written draft of the audit report is submitted first to the local plant manager, so that any differences can be resolved. It is then sent to the safety and environmental affairs vice-president for review, before it is formally issued. At this stage, if it appears that there are significant compliance shortcomings, the safety and environmental affairs vice-president may ask for a legal review.

It is important that out-of-compliance situations not be documented without a companion statement stating the company's planned efforts to remedy the problem. If it is subsequently decided not to carry out a recommendation, the reasons for this decision, and for any alternative course of action taken, must be documented and placed in the audit file.

Operating management is responsible for subsequent follow-up, to insure that recommendations are carried out in a timely manner. This is accomplished through a quarterly report, summarizing all of the audit recommendations outstanding, plus those completed during the last quarter. At the next audit for the facility, the previous audit recommendations are reviewed, and if still considered to be outstanding, are noted again on the new audit.

Summary

We have now been conducting these audits for three years. We feel that they have allowed us to anticipate problems, and that they do give us some degree of assurance that our safety and environmental responsibilities are being met. They have also given us an additional benefit, in that operating people who conduct the audits get to see how other plants operate, so that a feeling of unity of purpose develops in the organization. I believe that the audits are well worth the time and effort that has been invested in them.

Appendix A: Environmental Review Pre-visit Questionnaire

Organization

Please provide an organization chart for the facility, showing responsibilities of operating and staff personnel, form lead hands or crew leaders through to the facility manager. Show persons having responsibilities for air emissions, waste water discharges, waste handling and disposal, PCB storage, unexpected adverse effects, safety coordination, in-plant atmospheric air monitoring, emergency response coordination, underground tank monitoring, Red and Blue book maintenance, incident reporting, CAER program, and any other facility safety and environmental functions not specifically noted.

General

1. Describe any process modifications or expansions currently being planned.
2. Describe the major pollution control facilities employed at the site (manufacturing and laboratory).
3. List any areas of non-compliance associated with operations over the past year.
4. Highlight the major environmental concerns at the site, including any areas of regulatory uncertainty.

Air Emissions

1. List all substances routinely emitted to the air in measurable quantities, with annual quantities emitted.
2. What are the regulatory limits for these emissions (reference appropriate sections of the legislation).
3. Provide monitoring reports for the last two years.
4. What communications have there been with the Ministry over compliance or alleged violations?

Sewage Effluent

1. Describe the pretreatment process, if any.
2. List all substances routinely discharged in measurable quantities, with annual quantities discharged.
3. What are the regulatory limits for these discharges (reference appropriate sections of the legislation).
4. Provide monitoring reports for the last two years.
5. What communications have there been with the Ministry over compliance or alleged violations?

Waste Management

1. List all categories of waste generated, and tabulate for each category the quantity disposed of during the past two years, number of shipments, names of disposal contractors, and methods of ultimate disposal.
2. What communications have there been with the Ministry over compliance or alleged violations?
3. What quantity of waste, by category, is now in inventory?

Hazard Analysis

1. Please provide copies of any HAZOP analyses performed during the past two years.
2. List HAZOP recommendations not completed, with comments.
3. Please provide copies of insurance inspection reports received during the past two years, with relevant correspondence and comments on recommendations not completed.

____

____

(Location)

Appendix B: Environmental Review Checklist

Blue Book Adequacy

- Is all relevant federal, provincial and municipal safety and environmental legislation in the Blue Book?
- Does the facility follow a procedure to keep it up to date?
- Are there any suggestions to improve the Blue Book?

Red Book Adequacy

- What percent of all current facility emergency, safety and environmental procedures are in the Red Book?
- Are these procedures adequate?
- What additional procedures are lacking?
- What percent of relevant facility drawings are in the Red Book?
- What additional drawings are recommended?
- When was the last revision made to the facility Red Book?
- Are copies of the Red Book readily available to facility operating personnel?
- Are there any suggestions to improve the Red Book?

Ministry Orders, Program Approvals

- Are previous and current safety and environmental Ministry directives, orders, program approvals, registration certificates and permits filed in an organized manner, and available for ready reference?
- Are any of the above missing?

Ministry Permits for New Construction

- Are all government permits for new construction filed in an organized manner, and available for ready reference?
- Are any of the above missing?

Process Changes Since the Last Audit

- What process changes have there been since the last audit?
- Should any of these have suggested revisions to the Red Book?
- Should any of these have suggested a HAZOP?

Review Previous Year's Appropriation Requests for Strathcona: Industry Association Checklist

- Was the Strathcona Industry Association checklist used properly on all appropriate area responses?

New Raw Materials Introduced Since the Last Audit

- Were any new raw materials introduced to the facility since the last audit?
- Were all of these properly checked for safety and environmental hazards through the process formula transfer procedure?

Anticipated Process Changes

- Are any process changes anticipated during the next year?
- Should any of these suggest a HAZOP?

MSDS Files on Substances Handled at the Site

- How complete are the MSDS files for substances handled at the facility?
- Does each "work-site" (i.e., department) have an up-to-date file of each MSDS used at the work-site?
- Are there adequate procedures for insuring that these files are complete?

Review Incident Reports for Spills, Accidents, Emissions

- Are incident report files maintained in an organized manner?
- Are the incidents well documented and numbered?
- Are any files missing?

Review of Incident Reporting

- Do facility personnel have clear instructions on how to report incidents to the proper authorities in a timely manner?
- Were reports of incidents, where required by legislation, made to proper authorities in a timely manner?

Review of Imminent Hazards

- Are there any situations involving chemical production, use, storage or disposal that might be termed imminent hazards?
- Have these situations been subjected to a HAZOP review?
- Is work planned or underway to ameliorate any of these situations?

Emergency Response Capability—On-site and for Goods In-transit

- What emergencies have been anticipated?
- Are emergency response contingency procedures well documented and clearly understood by facility operating personnel?
- Are these procedures adequate?
- Are they reviewed and updated periodically by facility personnel?
- Have emergency response drills or exercises been held?
- Has employee training been adequate?
- Are training records maintained for at least three years?
- Is the in-transit response emergency supplies list adequate and up-to-date?
- Is there a "suitcase" packed for in-transit emergencies and are its contents and the other emergency response inventory segregated for use only in emergencies?

Bulk and T/L Incoming Materials Liability

- Is there an up-to-date list of bulk and T/L materials purchased FOB point of origin, with suppliers' twenty-four-hour emergency telephone numbers?

Status of Community Awareness and Emergency Response Program

- What is the current status of the CAER program?

Continuous Emission Monitors, Repairs, Adjustments, Calibration

- Are there any continuous emission monitors at the facility?
- Are they adequately maintained or provided with adequate backup?

Control of Non-point Sources

- Are there adequate controls on non-point potential sources of air or surface water pollution (loading and unloading operations, plant site run-off, in-plant transfer)?
- What is the likelihood of there being a reportable incident from these potential sources?

Spill Prevention Control

- Is equipment adequately designed and in place to minimize the chances of a reportable spill to the environment?
- Are there adequate written operating procedures for operations involving materials transfer, which will minimize chances of a reportable spill?
- Is there any possibility of groundwater contamination from past facility activities?

Underground Storage Tank Monitoring for Leakage

- How many underground storage tanks are there?
- What are their ages?

- Do they have cathodic protection? Active or passive?
- Are procedures in place to monitor underground storage tanks for leakage, according to S & E Standard No. 19?

Operation and Maintenance of Sewage Pretreatment Facilities

- Are sewage pretreatment facilities in good operating condition and well maintained?
- Are there adequate backup or spare parts in inventory to insure continuous operation?
- Are there adequate operating procedures to insure compliance?

Sewage Effluent Permits and Limits, with Record Keeping

- Is sewage effluent discharged in accordance with operating permits or municipal by-law?
- Were there any incidents on non-compliance in the last two years?
- Are records of tests adequate?

Waste Recycling Programs

- Is there an active program to recycle as much waste as possible?

Waste Management Generator and Handling Permits

- Are waste management generator permits in good order and readily accessible in one file?
- Are facility wastes properly categorized?

Waste Disposal Contracts

- Have waste disposal contracts been signed with each disposal contractor, in accordance with S & E Standard No. 24?
- Are current copies of waste handling permits and insurance certificates on file for each waste disposal contractor used during the past two years?

Waste Handling Procedures

- Is there an up-to-date waste management procedure at the facility, in conformance with S & E Standard No. 24?
- Is there a clear physical segregation of waste and rework material at the facility?
- Is all waste adequately and properly categorized and inventoried, showing date of waste "generation" (i.e., the date that it is decided that it is not be to reworked)?
- Is there a follow-up system to insure that waste does not become over-age without Ministry notification?
- Are waste storage areas neat and orderly, with waste clearly marked to show its category and date of generation?
- Is there a follow-up system for waybills to insure that copies are received from haulers with time limits?
- Is record keeping for waste handling adequate, with waybills and Ministry notifications in one central facility file?
- Has all waste shipped from the facility during the past two years been properly categorized and disposed of?

Review of Solid Waste Disposal Practices

- Has a leachate test been run on typical solid waste disposed of at the facility to insure that it can be disposed of safely?
- Has solid waste been properly disposed of during the past two years?

PCB Storage

- Are there any power transformers or power capacitors in use at the facility containing over 50 ppb PCBs?
- Are any PCBs stored at the facility?
- Are the storage areas well maintained in conformance with federal and provincial guidelines and regulations?

Environmental Audit of a Municipal Sewage Treatment Plant

M.G. Thorne

WHILE INDUSTRY DOES an environmental audit to ensure they meet sound environmental practices internally and externally, a municipality must ensure it meets the public's scrutiny of a safe environmental operation for its facilities. In the past when problems occurred that impacted the public—such as odours—we did not usually involve the public in the solution but just went ahead and built equipment or added processes to take care of the problem.

Now we have to prove to the public that we are doing a good job and not putting the community at risk. Our Highland Creek Treatment Plant (HCTP) was first constructed to process four million gallons per day to forty-eight million gallons daily, including thermal conditioning and incineration of sludge. In spite of enclosing most of our unit processes and treating the odour air, we still received numerous odour complaints during the spring and summer of each year. In 1987, after a meeting with local ratepayers, the Ministry of the Environment requested the Metropolitan Works Department to undertake "an engineering assessment of all the emissions and their environmental impacts individually and collectively in relation to the Environmental Protection Act." The Ministry of the Environment Central Region indicated that an audit was warranted as the concerns of the residents went beyond simple nuisance considerations to questions of health and there was a lack of test and monitoring data to respond adequately to these concerns, particularly emissions from the two-hundred-foot incinerator stack.

M.G. Thorne is Acting Deputy Commissioner, Works Department, Municipality of Metropolitan Toronto.

Scope

In September 1988, the Works Department retained Canviro Consultants, now CH2M Hill Engineering, in conjunction with Senes Consultant Ltd., to undertake an environmental audit of the facility. The overall objective of this environmental audit was to review and collate the data available relevant to the design, operation and performance of the Highland Creek Treatment Plant and to report on the findings as they relate to potential public health effects. Within this project scope, the following specific objectives were addressed:

1. evaluate the performance of the various unit operations at the plant and their compliance with relevant Ministry of the Environment (MOE) guidelines and with the Environmental Protection Act;
2. review the plant pollution abatement systems including the facilities for air treatment and odour control;
3. review the quality and quantity of ash from the incinerator, comment on the environmental impact of onsite ash storage and review alternative disposal practices;
4. summarize Metro's plans for plant expansion and process alterations and estimate any additional impact of these changes on the environment; and
5. critically examine the results of the studies conducted or supported by Metro at the HCTP and correlate the findings of these studies with the short-term and long-term health effects on the community.

Design and Operating Status

The design of the liquid and solid treatment trains at the HCTP generally conform to standard design practices. Operational conditions would be considered to represent Best Management Practice in most instances. Notable exceptions include:

- The aerated grit tanks are operating at long HRTs and low air flow rates. This, in conjunction with the long holding time for grit prior to disposal, may contribute to excessive odour generation from the headhouse.
- The operation of Porteous process contributes to excessive solids generation, overloading other solids handling processes in the plant.

As a result, waste activated sludge is stored in the primary clarifiers and contributes to odour problems in this area of the plant.

- Operation of the incinerators at a temperature of 620 degrees Celsius on the top hearth has been shown to be effective in complying with hydrocarbon emission standards. Minor instrumentation and operational improvements to the incinerators would improve the controllability of the combustion process.
- Plant-specific operating log sheets would facilitate monitoring of key operational and performance parameters.

Future plans to provide "firm capacity" for the liquid treatment train will reduce emissions from the facility by ensuring that the full treatment capacity is available during plant maintenance periods. Plans to substitute anaerobic digestion-dewatering-fluid bed incineration for the present sludge management process train will reduce odorous emissions associated with the operation of the Porteous process.

Performance and Compliance Status

Raw sewage quality at the HCTP is consistent with the original plant design in terms of conventional contaminants. Trace contaminant (metals and organics) concentrations, in most cases, are not substantially different from those in the raw wastewaters at other Ontario sewage treatment plants. Treatment of leachate at the HCTP does not adversely affect plant performance in the present quantities.

Liquid effluent from the HCTP has complied with MOE Policies 08-01 and 08-04 in terms of five-day biochemical oxygen demand and total suspended solids during the period 1986–1988. Three instances of marginal non-compliance with the effluent total phosphorus requirement have occurred during the same period.

Ash from the HCTP incinerator would not be classified as a hazardous waste of a registrable waste under Regulation 309. After issuance of MOE guidelines dealing with utilization of such materials, Metro should pursue alternative uses of the ash.

Based on the results of stack tests and the MAMU monitoring program, the HCTP is in compliance with all applicable offsite quantitative guidelines and standards under Regulation 308. Total hydrocarbon emissions at the point of discharge from the incinerator stack have been above the allowable limit specified in Regulation 308 on only one occasion under normal operating conditions. Two other excursions were related to testing of the incinerator to define the optimum operating point.

Odour Complaints

Odour complaints directed towards the HCTP have been registered by residents living in the vicinity of the plant since the late 1960s. The period 1971 to 1972 was a particularly odiferous year, during which numerous complaints were registered (25). Since that time, major modifications and expansions have been undertaken at the facility, including the Phase IV expansion and the installation of a multiple hearth sludge incineration system. As these changes have the potential to significantly alter the odour emission profile of the plant, the analysis of complaint data was limited to the period January 1986 to November 1988. It is felt that complaints registered during this period are most representative of the current odour-generating potential of the plant.

The period 1986 to 1988 also represents the time during which the Centennial Community Recreation Association (CCRA) promoted an organized effort to heighten residents' awareness of the plant and encouraged residents to officially register their odour complaints against the facility. The CCRA effort resulted in a significant increase in both the number of complaints and the detail with which complaints were recorded.

Under the CCRA program, residents were encouraged to report their odour complaints to one or more of the following institutions: CCRA, via monthly log sheets distributed to each house; Metropolitan Toronto (Metro) via telephone call to the HCTP; or to the Ministry of the Environment's Central Region Office or Spill Action Centre. In evaluating odour complaints data, all three sources were reviewed.

Over 1,900 individual complaints were registered during the three-year study period. All complaints were reviewed and consolidated to remove duplication arising from multiple reporting of odour incidents from individual residences. The resultant database used in the study consisted of approximately 1,500 odour complaints.

Based on 1988 data, there was a correlation between the direction of the receptor from the plant and the prevailing wind direction at the time of the complaint for approximately half of all complaints lodged against the Highland Creek Treatment Plant. This suggests that other odour sources may account for some or all of the balance of the complaints.

The odour incidents are reported primarily from a core of individuals, with 50 percent of complaints originating from approximately sixteen to twenty-five households. However, the large number of households reporting the balance of complaints indicates that odour incidents may be fairly widespread with odours detected by a large number of residents.

The majority of odour-incident days involved the reporting of three or fewer complaints. However, ten or more complaints were recorded on 13 percent of all incident days over the period 1986 to 1988. This suggests that odour incidents are not necessarily localized isolated events.

Furthermore, in comparison to other Metro TPs, the Highland Creek odour complaint record suggests that odour emissions from the plant are reported more frequently.

The potential for community odour complaints appears to be primarily dependent upon prevailing meteorological conditions and is not in large part due to any attribute process upset conditions at the Highland Creek plant. The dominant meteorological factor contributing to odour complaints is believed to be onshore winds. Odours are most likely to occur when wind speeds are light to moderate (i.e., less than 5.4 metres per second). The random distribution of incident days by day or week further substantiates the importance of meteorological conditions and discounts the impact of upsets or cycles in plant operations.

The temporal distribution of complaints parallels the diurnal lake/land-breeze cycle. Complaints begin in the midmorning hours as the onshore breeze develops and drop off rapidly in the mid-evening hours as offshore winds develop.

The dominant meteorological factor affecting offsite odour potential appears to be onshore winds with atmospheric stability playing a minor role. The absence of any clear correlation between plant operations and offsite odour complaints indicates that the plant may presently be a continuous odour emitter with the extent, duration and distribution of complaints dependent primarily on the prevailing meteorological conditions.

Risk Assessment

In recent years, residents living near the Highland Creek TP have shown a heightened awareness of the potential impact of the plant on their health and comfort. A first-order, screening-level assessment of potential health risks associated with emissions from the HCTP was undertaken to evaluate the incremental impact of the plant on nearby residents. The assessment is intended to provide order of magnitude estimates for the primary routes of exposure for the purpose of identifying situations possibly requiring more detailed investigation. In assessing the health risk, efforts were focused upon air emissions in general and heavy metals and organic compounds in particular.

Although it is acknowledged that emissions of biologically active agents (i.e., viruses and bacteria) may be associated with the plant, little data are available to quantify the rate of emission and types of biological agents present. Hence the potential impact of emissions of biologically active substance were not modelled in this screening level assessment.

In general, residents living near the plant may be exposed to emissions through various pathways (routes of exposure) including, for example, direct exposure via the inhalation of dust or gases, or indirect exposure from the ingestion of plants grown locally in contaminated soil.

Although detailed exposure pathway monitoring was beyond the scope of this study, screening level calculations were performed to provide order of magnitude indications of the potential significance of both the major direct and indirect exposure pathways.

To assess the risks, short-term and long-term exposure modelling was performed. The short-term modelling was used to evaluate the risk from direct inhalation during short-term exposure. The long-term modelling evaluated the risks associated with both direct and indirect exposure pathways.

The health risk assessment procedure used is based on the concept of Relative Margin of Safety (RMOS). The RMOS is the ratio of the calculated dose to a reference dose:

$$\text{RMOS} = \frac{\text{Calculated Dose}}{\text{Reference Dose}}$$

An RMOS value of 1 is associated with a lifetime risk of cancer no greater than one in a million, which is the reference risk level used in this study. Thus, an RMOS value greater than 1 indicates that the estimated exposure level exceeds the reference exposure level and therefore the risk to an individual exposed for his or her full lifetime would exceed 0.0000001 in the same proportion as the RMOS has to 1. Any RMOS value between 0.1 and 1 would indicate situations in which there may be some potential for the RMOS to exceed 1, although the potential health risks are likely to be small. Such RMOS values would identify situations where additional assessment of emission levels, exposure parameters and health evaluations criteria would be warranted. Where RMOS values are below 0.1, further consideration is not warranted.

Acid Gases. Acid gas group RMOS values range from 0.0003 to 0.008 with nitrous oxides accounting for approximately 75 percent of the total exposure for both child and adult receptors.

Metals. RMOS values based on metal emissions from the sludge incinerator range from 0.0001 to 0.005 for metals as a group with exposure to cadmium accounting for approximately 66 percent and 70 percent of the total exposure to adult and child receptors, respectively.

Dioxins/Furans. As no data were available on the composition of dioxin and furan emissions, RMOS estimates are based upon the most toxic congeners. RMOS values range from 0.00001 to 0.0005 with an approximately equal contribution from dioxins and furans to the total exposure.

Chlorinated Hydrocarbons. RMOS estimates range from 0.0001 to 0.007 with exposure to PCBs accounting for essentially all of the RMOS value for this group of chemicals.

Polyaromatic Hydrocarbons (PAHs). RMOS values for PAHs measured in stack-testing programs range from 1.7×10^{-11} to 1.0×10^{-9}. Phenanthrenes account for approximately 82 percent of the group RMOS for both the child and adult receptor.

Sewage Sludge Incinerator. The incremental contributions of the Highland Creek plant sludge incinerator to the RMOS are well below the level of 0.1 which is considered to represent an exposure of little consequence. The largest incremental RMOS contributions arise from emissions of acid gases, metals and chlorinated organics (primarily PCBs).

The total RMOS values for the acid gases, metals and chlorinated hydrocarbons groups are below 0.008 for both the child and adult receptor. Group RMOS values for dioxins/furans and PAHs are below 0.00005 and 0.00000001, respectively.

RMOS values calculated in the screening level assessment have not identified any potential concerns arising from the incremental risks resulting from the operation of the multiple-hearth sewage sludge incinerator.

These results are consistent with the findings of a 1987 MOE phytotoxicology soil survey in the vicinity of the WPCP.

Volatile Organics. Elevated group RMOS values result from three chemical compounds, chloroform, trichloroethylene and dichloromethane, with chloroform accounting for approximately 60 percent of the group RMOS. All other compounds assessed were well below the RMOS level of 0.1 which is considered to represent an exposure of little consequence.

Estimated group RMOS values for volatile organics are above 0.1 for adult receptors living within 1600 to 2000 metres from the plant and for child receptors living within 2000 to 2400 metres of the plant. Estimated RMOS values above 0.1 and below 1.0 indicate that there may be some potential for the actual RMOS to exceed 1, although the potential health risks are likely to be small. Group RMOS values approaching or exceeding 1.0, the level at which lifetime exposure will result in a lifetime cancer risk of one-in-a-million, have been estimated for a small fraction of the population (primarily northwest of the plant) living closer than 400 to 600 metres from the plant. Exposure at the estimated level will alter the base lifetime cancer risk level of one-in-a-million in the same proportion as the predicted RMOS is to 1.0.

In reviewing these RMOS values, it should be remarked that conservative assumptions were used throughout the analysis. Estimates of emissions of volatile organics assumed that the only available removal mechanism was volatilization to atmosphere and that air treatment equipment does not reduce the emission of these compounds. Estimates of exposure assumed that receptors spend their entire seventy-year lifetime within the influence of the treatment plant under unprotected conditions. Emission rate and RMOS values are therefore likely overestimated. When removal of volatile organics by biodegradation within the treatment plant is considered, the RMOS values for chloroform ranged between 0.1 and 0.95 for children and 0.1 to 0.6 for adults living near the plant. Although these levels are below the reference value of 1.0 and are still conservative, they are within the range for which further more detailed investigations are recommended.

Recommendations

The following recommendations were made based on the findings of the Environment Audit of the Highland Creek Treatment Plant:

- After issuance of MOE guidelines for utilization of non-registrable, non-hazardous solid wastes, Metro should investigate alternative uses of the incinerator ash.
- A sequential approach should be used to reduce odour emissions from the HCTP, to an acceptable level within a reasonable time frame. Initially, emphasis should be placed on the operation of the headhouse.
- A staged odour-monitoring program should be implemented to establish the plant's current emissions of odorous compounds and to monitor the effectiveness of in-plant odour control measures.
- A standardized odour-complaint recording system should be developed so that adequate data are available to support an odour monitoring program and to allow identification of odour sources and

development of additional control measures if required. A centralized registry system should be considered. The HCTP should be notified immediately of odour incidents so that the cause can be identified.

- A more proactive odour-surveillance program should be instituted at the plant with the knowledge that the HCTP is a potential source of objectionable odour emissions into the community.
- The meteorological station currently located on the roof of the administration building should be upgraded and relocated.
- Further quantification of emissions of volatile organics (e.g., chloroform) and their levels in local ambient air should be conducted.
- Revised HCTP monthly logs should be developed that are tailored to the specific configuration of the unit processes at the plant.

Audit Experience of a Middle-sized Oil and Gas Company

G.E. Younghusband

ON BEHALF OF the Independent Petroleum Association of Canada, I have been asked to make a short presentation on the real-world, first-time environmental audit experience of a mid-sized oil and gas company. What I intend to do in the fifteen minutes or so that I have been allotted is briefly describe the company I work for, explain my position with the company, tell you of our environmental policy, outline why it was felt an audit was necessary, cover the scope of the audit program and how it was organized, mention some of the problems we experienced and mistakes that we made, address lessons learned and give results in general terms, and finish with what we are doing with the audit reports.

I won't go into the history of the AEC Oil and Gas Company as many of you will be aware of how we got started. Suffice to say that we are a division of the Alberta Energy Company, a diversified corporation with interests in oil and gas, pipelines, forestry, coal mining and petrochemicals. The oil and gas company, of which Gwyn Morgan is President, is one of the top ten natural gas producers in Canada with sales of around 300 MMCF/D. The company is very active in exploration and has major gas developments underway in northern Alberta. In conventional oil and natural gas liquids, we produce in excess of 6,600 BPD. In heavy oil, we are a major owner of rights in the Primrose Range, and with joint venture partners, we are involved in a number of pilot plant extraction projects. In addition, AEC has a ten percent joint venture interest in Syncrude.

My title with the company is Senior Advisor, and among my many duties is responsibility for industrial safety and environmental protection. I have a small professional safety and range control staff under my direct

G.E. Younghusband is Senior Advisor, AEC Oil and Gas, Calgary.

supervision. The point I wish to make is that, unlike the majors, a small or mid-sized company like AEC Oil and Gas is unlikely to be able to afford the luxury of employing full-time professional environmentalists.

As a company, before you do anything in the environmental area you should have something to base your program on, that is, you need to state the company environmental policy, and this policy needs to be stated from the top.

It is all well and good to have a policy, but to ensure you are living by it you have to know where you stand. Most companies know in general terms if they have problems and can probably come up with a hit list of four or five items they really need to have a look at. This is where the audit comes in. In our case, the targets of our audit program were our thirteen compressor stations, four oil batteries and one sour gas sulphur extraction plant. Our aim was to quantify the problems in those areas where environmental protection performance did not meet the high standard articulated in our company policy statement.

Like most companies our size, we did not have the complete environmental expertise available in-house, and suitable people for separate environmental audits of eighteen sites were going to be hard to come by. This is where we made a key decision, which was a major factor in the success of our program.

We had already planned a safety audit of all our facilities so we combined the two programs and added an environmental consultant. The combined safety and environmental team was headed by our safety professional with team members, other than the consultant, coming from within the company. Individual team members changed throughout the program to ensure people were not involved in the audit of a site where they had some direct responsibility. We started with six team members, reduced to five and eventually to four as experience was gained. People were chosen for their experience in certain areas such as mechanical, electrical and production. While the environmental consultant was needed to set the guidelines, the team members rapidly picked up on situations relevant to their fields; for example, the mechanical members quickly began to identify problems and suggest solutions for disposal of lube oil, lube oil filters, barrels, oily rags and spill material; the production members were keenly interested in discussing clean-up techniques, disposal options for tank bottoms, treater clean outs and salt water spills; and the electrical members proved to be of vital importance as they are involved with systems that are the last line of defence, such as high and low-level alarms, gas monitoring and fire detection.

It is interesting to note that the line between safety and environmental concerns became virtually nonexistent as the team exercised their mandate to observe everything and anything relevant to their area of expertise. For instance, it was found that when inspecting bulk tanks located in a compressor yard, a team member would check for safety features such as Workplace Hazardous Material Information System and Transportation of Dangerous Goods markings, proper grounding cables, and then continue right into the area of spill protection by inspecting for improper wear on load hoses, proper nozzles with working shutoffs and proper beaming to contain accidental spills.

I should add, that although we tasked the teams with a combined safety and environmental audit program, we had the consultant produce formal audit reports on each site, reports which contained only items directly related to environmental protection.

The major problem we experienced and where we made a mistake in setting up and carrying out our audit program was in determining what we really wanted. As I mentioned earlier, we came up with a hit list of five items on which we wanted to place special emphasis. This was a good approach, but along with it we decided not to hit sites too hard on their first-ever environmental audit. Unfortunately, this sometimes led, in spite of our best intentions, to parts of reports which were too general in nature. Those old catch-alls like "improve employee awareness and training" frequently appeared, when we were looking for a good solid directive statement such as "clean up that spill area by removing contaminated topsoil, stockpile the soil and treat it, farm it or sent it to a disposal site." A second mistake we made initially was to form too large a team, however, this was corrected as the audit program progressed.

As for results of the audits, I do not intend to be site specific, but we obviously identified some areas where improvements were required. If we had not expected to find some problems we wouldn't have bothered with the audit. Some of the areas requiring attention, and I suggest areas that most oil and gas companies need to take a look at, were lack of tanks to contain washdown fluids and waste liquids, inadequate disposal procedures for empty barrels, compressor station site layouts, not taking into account the containment of spills, and the lack of proper run-off routes to prevent rainwater flushing contamination from some of the sites. In general terms, however, we were very pleased with the standard of environmental protection at most of our sites. As we should have expected, if your normal operating procedures are of a high standard you have already gone a long way toward ensuring you are providing a high standard of environmental protection. I would like to emphasize some other highly encouraging findings that came from our audits. Our field employees, and I am sure those of other oil and gas companies, are very aware of global environmental concerns like the pollution of air and water, the greenhouse effect, destruction of the rainforests, etc. They talk about these matters with their friends and families and they are very receptive when given the opportunity to participate and suggest measures designed to improve the standard of environmental protection at their work place.

I would like to say a few words on circulation of audit reports. We initially circulated the site reports to all levels of supervision for comment and direction on corrective action. Since one of the things we wanted to do was to move quickly on some of the findings, this broad circulation caused an unacceptable delay. We will, in future, restrict circulation of the reports to three levels of supervision: the foreman, the superintendent and the district manager. For higher levels of management, we will provide an executive summary.

Now that you have completed your audit program, what do you do with the reports? From our perspective there were basically two activities that were required as follow-up. The first is immediate follow-up when you discover a problem that must be addressed as quickly as possible. In this area we found it important to set time limits for action response. We also rapidly reinforced what all of us already know, that the "getting things done" activity is best handled at that level at which things are actually being done. By telling the field hands on site what needed to be done, letting them largely figure out how they were going to do it, and then supporting their proposals, we achieved some spectacular results, and usually at lower costs than we had anticipated.

The second action we are taking as a result of our audit program is what I would call long-term guidance to ensure AEC continues to provide the environmental protection expected from a good corporate citizen. We are using the audit reports to support the production of an environmental procedures manual and to provide direction on environmental protection for inclusion in our existing employee and contractor pamphlets. While a number of major companies have environmental procedure-type manuals, and the industry associations have provided some excellent environmental practices guidelines, and the federal and provincial government have published regulations and guidelines, our opinion is that most of these tend to be general and quite theoretical in nature, or they have been written for lawyers. As a consequence, although the industry publications provide excellent guidelines for management, most of what has been produced to date is of limited value in the field. We at AEC want a procedures manual that will be used, not left sitting on a shelf. The completion of our audit program has given us the background and specific knowledge to allow us to contract for the production of a manual which will provide single, clear direction to the people in the field.

Finally, I would like to finish by re-emphasizing what could be called advice based on the AEC audit experience:

- If you hire outside help from a consulting firm, which most of the smaller companies will need to do, make sure you are specific in what amount of detail you want in your reports. General motherhood type observations are not of much use.
- Don't aim too high on your first effort. You know or suspect your problem areas. Prioritize these, inspect them and develop corrective solutions.
- Consider combining safety and environment in one audit team. It worked very well for us.
- Set reasonable time limits for follow-up action, but set time limits, and follow-up to ensure action has taken place.
- Involve people in the field at all levels in the development of solutions to problems discussed. The improvements can be significant at a reasonable cost.
- Restrict the complete audit reports to those levels of supervision with a direct input; higher levels can be sent a summary of all reports.
- Have a plan for what you are going to do with the information gained from your audit program. It is going to cost you time, people and money, so get full value from your efforts.

I hope this presentation has been of some value. With the attention now being directed to all matters environmental, a company cannot afford to be found wanting. If you adopt a positive approach, spend some time thinking through what you want to look at and what you want to achieve, you will find that completion of an environmental audit and an aggressive effort to follow-up on the resulting recommendations, will allow you to face with some confidence impending legislative changes and what will most surely be the decade of the environment. At least that has certainly been our experience at AEC Oil and Gas.

Real Estate Preacquisition Audits

R.B. German

THE TERM "AUDIT," when applied to a pre-purchase environmental evaluation at a site, is generally taken out of context. An audit involves a process of verification—of checking a "paper trail" to ensure that any irregularities are identified—and is, in this case, normally undertaken to establish whether applicable environmental and related regulations are being met.

A "compliance audit," for example, involves a review of all legislative requirements which affect, either directly or indirectly, the operation or occupation of a property or building and a comparison of those requirements with records of and procedures for on-site activities. A "waste audit," on the other hand, is specific in nature and entails a program of determining whether all waste streams at a plant have been correctly classified, any necessary registrations have been completed, storage facilities and procedures follow regulatory requirements, shipping manifests are properly completed and distributed, appropriate disposal arrangements have been made and all necessary records are available and in order. Although audits of this nature are often undertaken prior to transfer of title of a property, they usually comprise part of an overall assessment associated with the sale of the industrial or commercial facility as a going concern.

When a plant is to be purchased with the intention of continued operation, the results of the various audits permit the prospective owners to assess different aspects of the operation of the facility. An audit does not, in most cases however, provide more than a component of the information necessary to identify, and ultimately ameliorate, the potential

R.B. German is Vice President of MacLarentech Inc., Willowdale, Ontario.

environmental liabilities associated with the reuse or redevelopment of an industrial property or of other lands contaminated during the course of past activities. It is the identification of these liabilities, through programs of site characterization, facility evaluation and compliance and operational audits, that must be addressed in completing a pre-purchase property assessment.

The presence of environmental contaminants or hazardous wastes and materials within a building or discharged to the ground surface during the course of past operations significantly affects the utility of the property. In order to achieve full and unencumbered use of the land and buildings, remedial action to resolve the outstanding environmental concerns would have to be implemented. Conversely, downgrading allowable land use to a level commensurate with existing environmental conditions could also be considered but may demonstrably effect the anticipated return to the vendor or the value of the purchaser's investment. Numerous examples of the effects of the presence of contaminants and controlled substances in soil and groundwater and in structures on sites across the country have come to light in recent years. The impact of these findings on the course of the transactions and on subsequent redevelopment effectively highlights the need for careful review of past land use, effective site characterization work, detailed facilities inspections and, where ongoing industrial activity is contemplated, operational and compliance audits.

Transactional evaluations are indispensable tools for both parties to a sale. It is in the interest of buyers to protect themselves from assuming unexpected environmental liabilities and of sellers to understand the extent of the financial burden that may have to be accepted in disposing of a property. Where an assessment is undertaken on behalf of a prospective purchaser, the environmental concerns that may come back to haunt a new owner after title is assumed, to have been enumerated. In effect, the consultant does his best to identify all possible reasons for not buying the site. When the vendor is the client, the consultant should establish the full extent of soil and groundwater contamination on-site and the presence of wastes and other hazardous substances to provide the owner with sufficient data to elect to decommission the facility prior to the sale, to obtain an appreciation of the liabilities that may be passed on to another party, or to retain ownership of the property.

A pre-purchase evaluation program should incorporate not only technical and regulatory components, it should also address factors affecting the value of the property as reflected in the cost of remediation and the reduction of off-site impact. These financial liabilities involve the extraordinary costs associated with the sale and redevelopment of the affected lands, including cleanup, achieving regulatory compliance, posting bonds and so forth and should be incorporated into, or noted with the appraised value of the property.

Risks

In preparing to undertake a pre-purchase evaluation, it is important to understand the nature of the environmental risks to which the purchaser of a potentially contaminated site can be exposed. These include:

- buried wastes—such as drums of chemicals, bulk solids, gas containers, off-spec product, foundry sand, etc.
- controlled substances—including PCBs in service in transformers, capacitors, and light ballasts as well as in licensed interim storage; asbestos fireproofing, insulation and building materials such as transite, floor and roof felts and tiles.
- underground storage tanks—for fuels, slops, overflow and process chemicals. It is interesting to note that United States Environment Protection Agency studies indicate that virtually all unprotected steel underground storage tanks in excess of five years of age exhibit evidence of leakage.
- contaminated soil, surface and groundwater—from process discharges, previous industrial activities on-site or nearby, fallout, the use of contaminated building materials, etc. This area of concern can have ramifications off-site as well as on the property.
- airborne contaminant plumes—stack discharges from an operating plant may exceed allowable limits and impinges on a receiver.
- process effluent discharge—are effluents or process chemicals entering the subsurface, sewers or ground and surface water receivers on- or off-site at levels exceeding allowable limits?
- waste management procedures—are wastes accumulating on-site, are they discharged or disposed of according to regulation?
- past land use—did the site or adjacent properties have previous industrial or other uses that may render it defective?
- soil gas accumulation—do radon, volatile hydrocarbons, methane or other gases enter or have the potential to enter and accumulate to deleterious levels in structures on-site?
- spills history—is there a history of non-compliance with environmental legislation that brings regulatory attention to the site?

Many of these risks are associated with the transfer of ownership of an industrial or commercial facility as a going concern and as such can, to some extent, be accommodated. They become critical, however, when a property is to be redeveloped for other more restrictive purposes such as residential or institutional. When a plant is purchased with the intent to continue operations, the buyer is generally cognizant of the exposure associated with the transaction. Operations can continue with a relatively high degree of on-site contamination without affecting the productivity or profitability of the facility until some potential change in land use or regulation is contemplated.

Developers, private individuals, municipalities, non-profit housing co-operatives, financial institutions, real estate investors and professionals such as lawyers, real estate brokers and so forth, on the other hand, are often relatively unsophisticated in an industrial sense. They have no reason to be aware of all of the risks that can be present, and so buyers and sellers often assume liability for situations that they inadvertently inherit.

Regulatory Issues

A plethora of environmental regulations and guidelines of varying degrees of severity and promulgated by numerous different federal, provincial, territorial and, in some cases, municipal authorities, exist in Canada. A compendium of those considered most applicable with respect to real estate transactions has been appended hereto and relates to:

- protection of the environment
- regulation of hazardous wastes
- control of liquid, solid and waste polychlorinated byphenyls and equipment
- management of asbestos and asbestos-containing materials
- transportation of dangerous goods
- control and management of underground and above-ground fuel and hazardous materials storage tanks
- water resources preservation
- decommissioning of industrial sites and cleanup of contaminated land
- disposal of contaminated soils, wastewater and other non-hazardous wastes
- occupational health and safety
- public involvement
- airborne emissions control
- environmental assessment.

Despite the range of legislative authority, however, government agencies are not generally in a position to force an owner to effect a cleanup of a property unless there are reasonable grounds to suspect that an offence or regulatory contravention has taken place. An owner, under current legislation, can essentially occupy contaminated land indefinitely, without embarking on any remedial program as long as:

- No deleterious materials in the groundwater or subsoil, at surface or in the air, are leaving or threaten to leave the property and impact upon any off-site receiver;
- Any spills of hazardous or other deleterious materials have been properly cleaned up and reported to the appropriate authorities;
- Hazardous wastes are not stored on the site for more than the prescribed time limit without an appropriate permit or as otherwise authorized under applicable legislation;
- PCB wastes are properly stored or disposed of in accordance with the regulations;
- Under- and above-ground fuel tanks have been removed or protected in accordance with the regulations;
- Wastes are disposed of in accordance with the regulations and unauthorized waste disposal facilities have not been established on the site;
- The property is secured from illicit entry and the public-at-large is protected from exposure to such contaminants as may be present on-site; and
- No change in land use is contemplated.

This is not generally the case in the United States where state and federal Environmental Protection Agencies have a much broader mandate to intervene than in Canada.

While it may appear as if the owner or purchaser of a property has broad latitude to avoid embarking on an evaluation or cleanup, it is obviously still incumbent upon him to ensure that site conditions comply with the above requirements, if only to provide evidence to the regulatory authorities that site conditions do not contravene any aspect of extant legislation. The purchaser and the vendor should also consider the advantages of establishing baseline environmental conditions relating to the plant structures; overburden, bedrock and groundwater regimes and in surface water bodies on the site at the time of sale. The results of the evaluation could provide the seller with a document indicating that conditions complied with regulatory requirements and guidelines at the time of the transaction and that any issues of an environmental nature raised in the future would thus have had to occur subsequent to sale. Conversely, the purchaser could structure the agreement between the parties in a way that assigned responsibility for future remedial action, as it relates to existing conditions at the time of transfer of title, to the vendor.

Financial Implications

The results of a detailed pre-purchase property assessment should provide the client with a representative estimate of the anticipated costs attendant on any environmental deficiencies associated with the subject of the transaction before an irrevocable commitment has been made. Depending on the condition of the site and facilities located upon it, a purchaser may find that he is required to incur a costly cleanup even to hold the site. Ultimate reuse of contaminated lands and plant cannot be assured without site characterization and remediation.

The financial implications of decommissioning a site are usually significant. The costs associated with deactivating a plant and implementing a remedial program, whether by the vendor or the purchaser, can be very high, particularly for an old facility which may have operated for many years with little or no regulatory control or guidance.

The environmental status of a property can have a direct affect on land values. A piece of land warranted "clean" by a seller, and thus suitable for redevelopment, will be appreciably more valuable than land known, or believed, to contain contaminants and wastes and for which no certainty exists with respect to the receipt of approvals for redevelopment.

Where a purchaser assumes ownership of a piece of property or facility found, after the fact, to be contaminated, he or she can be subject to a wide range of direct costs and losses due to:

- downgrading of allowable land use—residential usage, for example, may be proscribed because of contaminant levels;
- reduction of intrinsic worth—the extent of contamination on-site may be found to bring the realizable value of the property to well below the purchase price;
- direct cost of cleanup—out-of-pocket costs not anticipated at the time of purchase can amount to many millions of dollars for soil and groundwater cleanup;
- civil action—the costs of legal defence and of assessed damages resulting from court actions brought by aggrieved purchasers or

adjacent property owners impacted by contaminants leaving a newly acquired site;

- disposal—disposal costs associated with removing waste deposited on a property by others will be assumed by the purchaser;
- facility deactivation/decommissioning—the costs of full-scale decommissioning of a plant and site are significantly greater than those associated with simple demolition.

The costs associated with an unexpected cleanup, responding to an order issued by a government official, in defending oneself in court against a suit or charge or of working under the requirements of a mandatory injunction can be considerably higher than the ultimate return anticipated from the property itself. In addition, of course, once an unexpected environmental problem has been identified, the owner may not be in a position to easily dispose of the site to regain any of the initial investment.

In Ontario, and in most other jurisdictions in Canada, it is the owner of record who is responsible for chronic environmental issues on a property, although a lessor or operator of a facility on land owned by others must comply with all applicable laws and regulations, including those relating to discharge to and contamination of the environment. To be sure, a purchaser has recourse through the courts for recompense and can involve all involved parties including lenders, appraisers, agents, solicitors, consultants and so forth. Once a new owner finds that he has fallen heir to an unexpected problem, costs can mount quickly. In Quebec, however, the purchaser is always fully responsible for cleanup.

Pre-purchase Environmental Evaluations

The environmental studies carried out prior to a real estate transaction should be considered as integral parts of the investment valuation process, on an equal footing with property appraisal, the status of liens and titles, ability to insure, marketing issues, returns from plant production, etc. As with any primary consideration affecting a significant investment, all concerns of an environmental nature should be identified and resolved at an early stage of the property or facility purchase process. Unfortunately, pre-purchase evaluations relating to environmental issues are still generally requested as an afterthought, often with only a few days remaining before closing.

Pre-purchase studies are best conducted on a staged basis to allow the data generated during the course of each phase of the process to be evaluated, and, where appropriate, used in planning subsequent investigative or remedial activities. The wide range of environmental issues that affect the utility and value of a property may require that a number of individual studies be undertaken to permit an adequate assessment of each concern to be effected. Given that some degree of clarification, undertaking or cleanup, up to and including full-scale decommissioning and demolition, may be called for before transfer of title, an inadequate time frame for completion of the necessary work can result in inordinate delays to closing, resulting in increased interest costs, and could perhaps jeopardize the entire sale. The process should thus be instituted immediately following the preparation of an agreement to purchase, which should allow a sufficient period of time for either, or both, parties to the transaction to undertake the necessary environmental studies.

A site evaluation is an iterative process with the results of each successive step often dictating the nature and extent of subsequent studies. These steps include:

- a preliminary site reconnaissance
- a historical review of land use and ownership
- discussions with the regulatory authorities relating to plant or site concerns
- facility evaluations relating to buildings, plant and other structures on site, including a compliance audit
- site characterization (subsurface) investigations relating to soil, bedrock and groundwater conditions
- identification of applicable environmental liabilities
- recommendations relating to supplementary investigation and remediation requirements
- preparation of preliminary cost estimates for any necessary additional studies and remedial action.

The findings of pre-purchase investigations should be sufficiently comprehensive to permit their use in preparation of detailed work plans for subsequent studies and remedial design programs where the need for cleanup or decommissioning work is evident. To avoid duplication of effort, the environmental work undertaken prior to a transaction should not be completed in isolation. It must be developed with the full understanding of the issues relating to remediation and in such a manner as to facilitate incorporation into supplementary studies, should they prove to be necessary.

The findings of a pre-purchase evaluation can have a direct impact on the interest of all parties to the transaction, including the vendor and purchaser, the lending institutions, legal counsel, real estate brokers and so forth. Since the compliance of the property owner, and of a lessor where a tenant is involved, would be required before a prospective buyer implements an investigation, it is only reasonable to provide the findings of the studies for their review and comment. Similarly, the results of site and facility evaluations undertaken by the vendor and of any subsequent remedial, verification or decommissioning reports identifying the extent of any cleanup work undertaken and reports of any known residual latent defects should be provided to the purchaser. The nature of any remaining liabilities can then be properly understood and quantified.

The mechanism for undertaking, and transferring the results of, the various environmental studies included such issues as access, restrictions because of process secrets, discussions with regulatory authorities and plant staff should be detailed in a mutually acceptable manner in the purchase agreement. It is also important that the wording of the agreement properly define the terms used so that no ambiguities arise in the interpretation of the intent of the document with respect to the right of inspection and the relevance of the findings of the evaluations. Such general phrases as "items of environmental concern" and "environmental contaminants" are sufficiently vague to permit a wide range of meanings to be assumed by either party to the transaction.

Site Reconnaissance

A preliminary site visit, which should include a walk-over of the property and a tour of any buildings or plant facilities on the site, constitutes the first stage of a pre-purchase evaluation program. At this point, the work should be limited to discussions with staff, an initial review of site plans, flow sheets and plant drawings and a visual appraisal of the general condition of the land, any processes or industrial operations and buildings. A clean, orderly and well-maintained site facilitates the on-site inspection work at this and at all later stages of the environmental evaluation. It also generally eliminates visible indicators of concern that garner the interest of regulatory inspectors.

The tour of the property and facilities should include:

- a boundary-to-boundary walk across the property
- sewer and surface drainage lines and septic systems

- surface water bodies and water or waste treatment facilities
- on-site disposal or fill areas
- underground and above-ground storage tanks
- pipelines and pumping facilities
- waste storage areas and facilities
- plant roadways, haulage routes, parking and laydown areas
- lagoons and reservoirs
- wells and water supply systems
- shipping and receiving areas
- process areas
- pits and chambers
- hazardous materials and products storage
- maintenance and repair areas
- QA/QC, development and analytical laboratories
- pilot plants
- boiler rooms and heating facilities
- air pollution control systems including scrubbers, baghouses, etc.
- electrical substations, distribution and lighting systems and machinery controllers
- steam and hot water distribution
- refractory systems
- containment and spill control
- areas of staining and discharge
- drains and sumps
- surrounding land use.

The information obtained during this phase of the study, which is often undertaken as part of the proposal preparation process by the consultant, is used to define the scope of the evaluation and to establish or amend initial estimates for the work. It is at this stage that a basic understanding of the range of environmental liabilities associated with the property can first be appreciated.

Historical Review

The history of a site provides an unique perspective on the facility and property subject to pre-purchase assessment. Past activities on and in the vicinity of, a parcel of land subject to sale can have a significant effect on its environmental condition. Experience with sites like McClure Crescent, Lees Avenue and a number of other well-publicized "Love Canals" have shown only too well how latent environmental defects can affect developed property values and construction activities and costs.

The historical review process permits the identification of areas of concern on a site through an evaluation of past land use and ownership. A typical historical search involves a title search through the local land registry office and archive searches of old municipal directory data and fire insurance maps. Old maps and photographs of urban areas were often annotated with information relating to land use and can thus also be of use in identifying areas of concern. A review of sequential airphoto data can also be of significant value in identifying activities on a site, particularly with respect to building extensions, landfilling and excavation work and to waste burial and bulk materials and waste storage. Finally, interviews with long-term and retired employees, nearby residents, and municipal officials can provide a wealth of data relating to specific events or concerns such as spill and dump or discharge sites, the location of old underground storage tanks, drum burial areas and the like.

The compilation of all relevant data relating to the site provides both general information with respect to the nature of any contaminants associated with past industrial, disposal or other activity and specific direction on the location of buried wastes, the sites of previously removed storage tanks, maintenance areas, etc. The juxtaposition of historical information with current site plan data permits the development of a comprehensive site investigation and facility inspection process which is properly directed at the relevant issues on site.

Regulatory Review

The review of regulatory issues affecting a site involves, in part, the completion of a compliance audit to determine whether all necessary records, licenses, permits, certificates of approval, registrations, inventories and so forth are in order. While many of these requirements are applicable only if continued operation of a plant is considered, a number are transferrable with title and become the responsibility of the purchaser. Examples include:

- air pollution control
- hazardous waste registrations
- waste manifesting
- approved PCB storage facility
- inventory of PCB equipment and wastes

- inventory of asbestos-containing materials
- inventory of underground and above ground storage tanks
- special discharge permits for sewer by-law exceedences
- waste storage period extensions
- notification and removal of out-of-service underground storage tanks
- permits to take water
- maintenance of updated Material Safety Data Sheet files
- waste disposal facility
- hazardous goods and waste haulage by licensed carriers
- placarding in accordance with Transfer of Dangerous Goods requirements
- waste disposal at approved sites.

These requirements should be available on site or through corporate sources, if records and personnel are in place.

The regulatory review process also involves the perusal of records relating to outstanding or past charges or orders laid by provincial or territorial environment departments with respect to environmental protection act or regulation infringements, by municipal works staff relating to sewer discharge violations, as well as to public complaints, lawsuits or other legal actions. Local or district environment ministry office personnel will normally provide all necessary information relating to complaints and orders, although permission to obtain access to the files in some jurisdictions, such as Alberta, may be required from the vendor.

Information relating to past regulatory violations can provide an appreciation of the probable condition of the property and of the structures located on it. If the site has proven to be of substantial concern to the authorities in the past, it is probable that the purchasers will become the subject of scrutiny of assuming title. Further, that attention of this nature at the site could affect the ease with which approvals for cleanup or disposal of wastes from the property can be obtained.

Preparatory Activities

Prior to commencing with the field investigations stage of a pre-purchase environmental evaluation program, a series of preliminary undertakings would be required to:

1. provide a base plan for facility and layout drawings suitable for site investigation and facility evaluation purposes, as well as for future

remedial or decommissioning program design, tendering, contaminant distribution, verification testing and as-built purposes;

2. prepare a health and safety plan which reflects the full range of plant and site hazards; identifies the appropriate level of personal protection; details exposure limits, symptoms and recognition qualities, first aid and monitoring requirements; outlines safe work practices and contamination control procedures and provides a detailed listing of local emergency contacts;
3. develop a detailed work plan that identifies targets for inspection, locates all proposed boreholes, monitoring wells, piezometers, sample stations and surficial sampling locations; identifies equipment requirements and details investigation procedures; outlines air and other real time monitoring procedures; enumerates field protocols for sampling, sample handling, documentation and chain of custody; establishes minimum requirements for decontamination of equipment; identifies lab-testing requirements and quality assurance/quality control protocols and provides plant or site survey grid, baseline locations and benchmark data for field control purposes; and
4. clear municipal and plant services including sewer, watermain, electrical, gas, telephone, cable and other utilities to avoid disruption of service during subsurface investigations phases of the site characterization work.

Where possible, the use of a digital mapping base and a geographical informations systems to prepare base plans and working drawings is recommended for flexibility and the facility of interfacing with computer assisted drafting design equipment for the preparation of design drawings, sections, quantity take-offs and so forth.

Facility Evaluations

The performance of a facility audit as part of an overall environmental assessment represents a primary component of any property evaluation. The buildings and/or processes contained therein may contain or utilize any number of chemical substances deemed to be environmentally hazardous. Specific environmental or occupational health and safety regulations may also have been enacted for the control of usage or disposal of these substances. In addition to the potential hazards to the environment, workers on-site and the general population, the presence of such materials may impose an environmental liability upon the property which becomes the responsibility of the owner of record. This environmental liability is of particular concern when the structures are scheduled for demolition or major renovation as a function of redevelopment of the property.

In situations where a change of ownership occurs, the purchaser will assume liability for all hazardous materials remaining on-site, whether or not the materials pose an immediate or potential hazard. While the impact of this liability may be deferred if the facility is purchased as a going concern, contaminated materials and controlled substances management programs will have to be implemented as soon as any form of redevelopment or renovation is contemplated. The costs for clean up of contamination resulting from previous land use and for the removal of controlled substances such as asbestos, polychlorinated biphenyls (PCBs) or hazardous raw materials, products or wastes can and, in many cases does, contribute significantly to the overall cost of purchase and development of the property.

Controlled substances of significant concern include asbestos and PCBs. These two materials were used extensively for a period of forty to fifty years prior to 1980 and are usually present in most structures built before 1980. Other materials that cause concern include any hazardous chemicals utilized in specific manufacturing processes and the resultant waste streams. The liability imposed by the presence of hazardous compounds and materials will also depend on the measured degree of soil and groundwater contamination associated with their presence or with past usage on-site, the quantities used, the methods of containment, clean up and disposal utilized during plant operations.

Although underground storage tanks are generally inspected as part of the facility evaluation component of a pre-purchase evaluation, the impact of their presence is best assessed during the course of the investigations undertaken as part of a site characterization program. That being said, however, the identification of tank locations, review of fuel or product storage, receipt and dispenser inventories, and a listing of the materials and quantities in storage comprises an integral part of the facility evaluation work undertaken at a property prior to the transaction.

The evaluation process involves a careful and systematic inspection of all aspects of the plant building, storage facilities, products, wastes, processes and operational activities associated with the property. A comprehensive checklist is adapted from a standard master form to reflect the specific facility concerns identified during the initial reconnaissance and is used by the inspectors as a guide during the course of the site work.

The primary components of an inspection include:

PCBS

- transformers
- capacitors
- switch gear
- fluorescent light ballasts
- ballasts
- electro-magnets
- hydraulic systems
- accumulated oils
- licensed interim storage facilities
- registration with Environment Canada.

Asbestos Materials—Fireproofing

- pipe insulation
- insulation for boilers, heat exchangers, hot water tanks and ducts
- thermal insulation
- acoustic barrier
- gaskets and mastic materials
- roof and floor felts
- ceiling and floor tiles
- "transite" construction products, including pipe/siding materials.

Storage Tanks

- underground and aboveground
- proprietary (third-party ownership)
- cryogenic and heated tanks
- dispensers, piping, vents and fillers
- spills evidence
- pressure tests
- residual levels
- product identification.

Hazardous Goods

- raw materials
- products
- drummed and bulked hazardous wastes
- accumulated wastes in sumps, pits and drains
- storage facilities

- ventilation, containment and fire control
- identification of unknown wastes
- registration and disposal
- product information and Management Systems Data Sheets
- licensed haulers used
- periods of accumulation
- labelling and inventory control
- contingency response
- spills control
- storage drum, carboy, bag and carton conditions
- general housekeeping
- licensed and unlicensed waste disposal facilities.

Air Emissions

- certificates of approval
- baghouses and filters
- scrubbers
- stack condition
- fume control
- particulate and vapour discharge.

Wastewater

- storm and sanitary sewer discharge
- condition of sewers, drains and septic systems
- waste accumulations in pipes, lines, leaching beds and sumps
- solids control
- spill containment and drain protection
- water treatment plant.

Process Review

- process description and flow sheet
- mass balance
- condition of equipment
- age and state of repair
- discharges and leaks
- housekeeping and maintenance
- workplace hazardous material information system preparedness, health and safety and training
- ventilation.

The findings, analytical results and observations of the facility assessment are presented in detailed report form and, where appropriate information is gleaned during the course of the inspection, used to refine the site-characterization program proposed for the property.

Site Characterization

Subsurface investigation work is carried out during the course of a prepurchase evaluation to provide information on soil, bedrock and groundwater conditions beneath the site, on surface water conditions in any watercourse, ponds or lakes and on the distribution of contaminants and wastes, if any, across the property. This phase of the program is intended to determine the extent, nature and loading of the inorganic and organic compounds present and to establish whether they exceed the limits beyond which some form of remedial action may be deemed necessary.

Inasmuch as the object of a property transaction on the part of the purchaser generally involves a speculative investment, some form of redevelopment or continued operation of industrial or commercial facilities on the site, the field investigations are required, as a minimum, to determine whether:

- such contaminants as may be present exceed allowable redevelopment guidelines for industrial land, necessitating remedial action to effect a clean up
- groundwater degradation has occurred to the extent that recovery and treatment is warranted or whether special handling and discharge procedures need to be implemented during the course of any construction, remedial or decommissioning programs on site
- contaminants are fluxing across property boundaries to impact upon neighbouring receivers
- buried hazardous wastes or leachate toxic soils, requiring disposal at a secure landfill, are present
- soil can be left in place on the site, can be used as general backfill or requires disposal at a licensed landfill facility.

A relatively limited range of subsurface investigation techniques are available for site characterization work. These include both indirect and direct methods such as electromagnetic, ground-probing radar and seismic geophysical surveys, soil/gas surveys, test pitting, drilling and sampling and hydrogeological testing and can generally be implemented on relatively short notice.

In undertaking any investigation program for environmental purposes, the possibility of cross-contamination between samples and boreholes is always present. Careful sampler and, in many cases, drilling tool cleaning protocols must therefore be adopted to ensure that the analytical results are representative of actual soil and groundwater conditions. By the same rationale, no drilling fluids or additives may be used in the drilling process. The use of dedicated groundwater samplers, permanently installed in each monitoring well and discharged through a disposable filter to remove suspended solids, also constitutes good field practice.

Boreholes are usually advanced with the use of a mobile power auger rig, equipped with hollow-stem auger equipment for drilling in overburden and with diamond coring equipment for advancing boreholes into rock. Soil samples are typically recovered with the use of a standard 600 mm long, 55 mm diameter stainless steel split-spoon sampler in conjunction with the Standard Penetration Test. Samples are initially recovered at 750 mm intervals to intermediate depths of from 4.0 to 4.5 m and at 1500 mm intervals thereafter. Surface samples, and those recovered from test pits excavated by hydraulic backhoe, are normally recovered manually with the use of hand auger equipment.

Groundwater samples for chemical analysis purposes and water levels for establishing hydraulic gradients across the property, are usually recovered in monitoring wells installed in selected boreholes. The wells normally comprise slotted screens connected to "Triloc" riser pipes and set into an appropriately sized gravel pack of inert silica sand. The screen is sealed from the surface with a bentonite plug and the well casing is then grouted into place for security purposes. Where a monitoring well is to be used for sampling purposes, a "Waterra" pump is installed for well-development work and sample recovery.

Laboratory-testing work on soil, groundwater and waste samples should always be carried out using the standard procedures accepted by the regulatory authorities in the jurisdiction in which the property is located. This approach allows for reproducibility of results when work is carried out by different parties and ensures that confusion cannot arise during review of different sets of analytical results.

The actual range of testing conducted on samples recovered from a site would depend on historical land use and on the industrial, contaminant discharge or disposal activities that have occurred up to the present. Typically, testing work would be carried out for the following classes of inorganic and organic chemicals:

- basic surface and groundwater chemistry including heavy metals, inorganic parameters
- halogenated volatile compounds
- non-halogenated volatile compounds
- Water-soluble volatile compounds
- base-neutral extractable compounds (including polycyclic aromatic hydrocarbons—PAHs)
- phenolics
- PCBs and neutral chlorinated extractable compounds
- pesticides and herbicides
- Schedule 4 leachate test for metals, organics, anions, and PCB (Ontario)
- decommissioning guidelines (phytotoxicological parameters amended with selected organic compounds)
- lakefill guidelines (Ontario)
- sewer discharge by-law parameters

The extent to which some form of decommissioning or cleanup work will have to be implemented at a site is established on the basis of a review of the analytical results obtained with established guidelines or standards established by the regulatory authorities.

A detailed report on geological, hydrogeological and contaminant conditions at the site is prepared following review of the findings of the site-characterization program. It would present information with respect to the distribution of contamination and the scope of any necessary soil or groundwater cleanup and should include disposal options.

Liabilities, Estimates and Recommendations

Once all of the information generated during the course of the pre-purchase investigation program has been reviewed and assessed, the nature of the environmental liabilities subject to transfer with title can be established and quantified, in general terms, on a cost-equivalent basis. The costs developed would reflect the value of the remedial or decommissioning work necessary to permit redevelopment to proceed as well as estimated charges for additional investigation, design or related studies considered to be necessary at a later date. They would not, however, include estimates of fines or assessed damages.

In addition to the assessment of liabilities, recommendations relating to regulatory compliance issues, additional investigations and remedial requirements covering such concerns as underground storage tank removal and cleanup, buried drum removal, asbestos management and so forth would be prepared to provide the client, whether purchaser or vendor, with the information necessary to identify the extraordinary costs associated with environmental risks and remediation.

Confidentiality

The issue of confidentiality regularly arises in discussions with purchasers and vendors, and usually relates to the consultant's duty to disclose information obtained during the course of a facility evaluation or site characterization. It is, in our opinion, quite clear. The consultant's responsibility is initially to the client. Thus, when any violations of the environmental regulations are noted, the consultant is obligated to bring the issue to the attention of his or her client and recommend compliance. If repeated requests for compliance or redress go unheeded, the consultant's next approach should be to the client's solicitors with a request for assistance in resolving the matter. Only after fully exhausting all other avenues would the necessity of approaching the regulators be considered.

Appendix: Regulations/Legislation Pertinent to Facility Audits

Canada

Canadian Environmental Protection Act, 1988

Alberta

Clean Air Act RSA 1980, C. c-12
Clean Air Act: Clean Air Regulations Alta. Reg. 33/73
Clean Air Act: Clean Air (General) Regulations Alta. Reg. 216/75
Clean Air Act: Clean Air (Maximum Levels) Regulations Alta. Reg. 218/75
Clean Air Act: application For Approval of Natural Gas Compressors ERCB IL-80-30 Dec/80
Clean Water Act RSA 1980, c. C-13
Clean Water Act: Clean Water (Authority Designation) Regulations Alta. Reg. 200/77
Clean Water Act: Clean Water (General) Regulations Alta. Reg. 35/73
Clean Water Act: Clean Water (Industrial Plants) Regulations Alta. Reg. 36/73
Department of The Environment Act RSA 1980, c. D-19
Stop Order Appeal Regulation Alta. Reg. 9/74
Hazardous Chemicals Act RSA 1980, c. H-3
Hazardous Chemicals Act: Hazardous Waste Regulation Alta. Reg. 505/87
Transportation of Dangerous Goods Control Act SA 1982, c. T6.5
Transportation of Dangerous Goods Control Regulation Alta. Reg. 383/85
Transportation of Dangerous Goods Control Act: Administry Regulation Alta. Reg. 37/86
Water Resources Act RSA 1980, c. W-5

British Columbia

Cowichan River Estuary Environmental Assessment Procedures BC Reg. 486/77
Fraser River Estuary Environmental Assessment Guidelines
Fraser River Estuary Environmental Assessment Procedures BC Reg. 202/77
Environment Management Act SBC 1981, c. 14
Environmental Impact Assessment Regulation OC 1752/81
Environmental Appeal Board Procedure Regulation BC Reg. 1/82
Sanitary Regulations BC Reg. 142/59
Pollution Control Objectives for Food—Processing, Agriculturally Oriented, and Other Miscellaneous Industries
Pollution Control Objectives for Municipal Type Waste Discharges in British Columbia
Transport of Dangerous Goods Regulation BC Reg. 103/85
Transport of Dangerous Goods Act SBC 1985, c. 17
Waste Management Act SBC 1982, c. 41
Waste Management Act: Instructions for Applying for a Permit
Special Waste Regulations BC Reg. 63/88
Criteria for Managing Contaminated Sites in British Columbia, 1988

Manitoba

The Environment Act S.M. 1987–88, c. 26
Classes of Development Regulation Man. Reg. 164/88
Guidelines for Various Air Pollutants (Manitoba Department of Mines, Resources and Environmental Management, January 1978)
Hazardous Materials Designation Regulation Man. Reg. 15/81
Licensing Produres Regulation Man. Reg. 163/88
Private Sewage Disposal Systems and Privies Regulation Man. Reg. 95/88R
Storage and Handling of Gasoline and Associated Products Regulations Man. Reg. 97/88 R
Dangerous Goods Handling and Transportation Act CCSM c. D12
Classification Criteria for Products, Substances and Organisms Regulation Man. Reg. 282/87
Dangerous Goods Handling and Transportation Act: Environmental Accident Reporting Regulation Man. Reg. 439/87

Dangerous Goods Handling and Transportation Act: Federal Regulations Adoption Regulation Man. Reg. 172/85
Dangerous Goods Handling and Transportation Act: Generator Registration and Carrier Licencing Regulation Man. Reg. 175/87
Dangerous Goods Handling and Transportation Act: Manifest Regulation Man. Reg. 139/88
Dangerous Goods Handling and Transportation Act: PCB Storage Site Regulation Man. Reg. 474/88
Manitoba Hazardous Waste Management Corporation Act CCSM c. H15
Protection of Water Sources Regulation Man. Reg. 326/88 R
Rivers and Streams Act
Transboundary Pollution Reciprocal Access Act

New Brunswick

Clean Environment Act RSNB 1973 c. C-6
Air Quality Regulation NB Reg. 83-208
Clean Environment Act: Appeal Regulation NB Reg. 84-179
Clean Environment Act: Environmental Impact Assessment Regulation NB Reg. 87-83
Clean Environment Act: Water Quality Regulation NB Reg. 82-126
Petroleum Product Storage and Handling Regulation NB Reg. 87-97
Transportation of Dangerous Goods Act SNB 1988c T-11.01
Transportation of Dangerous Goods Act: NB 89-67

Newfoundland

Dangerous Goods Transportation Act SN 1982 c. 45
Dangerous Goods Transportation Regulations Nfld. 304/85
Department of the Environment Act SN 1981 c. 10
Department of Consumer Affairs and Environment Act: Air Pollution Control Regulations Nfld. Reg. 26/81
Department of Consumer Affairs and Environment Act: Environmental Control (Water and Sewage) Regulations Nfld. Reg. 156/80
Department of the Environment Act Storage and Handling of Gasoline and Associated Products Regulations Nfld. Reg. 258/82
Environmental Assessment Act: Environmental Assessment Regulations Nfld. Reg. 225/84

Department of the Environment and Lands Act: Mobile PCB Destruction Facility Regulations Nfld. Reg. 102/89
Waste Material Disposal Act SN 1973 No. 82
Waste Material (Disposal Act), 1973: Storage of PCB Wastes Regulations Nfld. Reg. 230/88

Northwest Territories

Environmental Protection Ordinance RONWT 1974 c. E-3
Transportation of Dangerous Goods Act SNWT 1982 (3rd sess.) c. 19
Transportation of Dangerous Goods Regulations NWT Reg. 048-85

Nova Scotia

Dangerous Goods and Haxardous Waste Management Act SNS 1986, c. 7
Asbestos Waste Regulations NS Reg. 238/88
Dangerous Goods and Hazardous Wastes Management Act: PCB Storage Regulations NS Reg. 225/88
Dangerous Goods and Hazardous Wastes Management Act: Dangerous Goods Storage Regulations NS Reg. 97/89
Dangerous Goods and Hazardous Wastes Management Act: Petroleum Storage Regulations NS Reg. 33/88
Dangerous Goods and Hazardous Wastes Management Act: Waste Oil Regulations NS Reg. 212/88
Environmental Assessment Act SNS 1988, c. 11
Environmental Assessment Regulations NS Reg. 94/89
Environmental Protection Act SNS 1973, c. 6
Environmental Protection Act: General (Suspension or Cancellation of a Permit) Regulations NS Reg. 221/86
Environmental Protection Act: Suspension or Cancellation of a Permit Regulations NS Reg. 116/86
Water Quality Standards (Suspension or Cancellation of Permit) Regulations NS Reg. 74/86
An Act Respecting the Transportation of Dangerous Goods SNS 1982, c. 5
Dangerous Goods Transportation Act General Regulations NS Reg. 152/85
Water Act RSNS 1967, c. 335

Ontario

Dangerous Goods Transportation Act, 1981 SO 1981, c. 69

Dangerous Goods Transportation Act: General Regulation O. Reg. 460/89

Energy Act: Fuel Oil Code RRO 1980, Reg. 288

Environmental Protection Act. O. Reg. 435/89

Environmental Protection Act RSO 1980, c. 141

Environmental Protection Act: Effluent Monitoring—Inorganic Chemical Sector O. Reg. 359/89

Environmental Protection Act: Air Pollution Control (General) Regulations RRO 1980, Reg. 308

Environmental Protection Act: Ambient Air Quality Criteria Regulation RRO 1980, Reg. 296

Environmental Protection Act: Boilers Regulation O. Reg. 16/86

Environmental Protection Act: Effluent Monitoring—General O. Reg. 695/88

Environmental Protection Act: Spills Regulation O. Reg. 618/85

Environmental Protection Act: Waste Management Regulation RRO 1980 O. Reg. 309

Environmental Protection Act: Waste Management PCB Regulation O. Reg. 11/82

Environmental Protection Act: Guidelines for the Treatment and Disposal of Liquid Industrial Waste in Ontario Dec. 1978

Environmental Protection Act: Guidelines for Protection Measures at Chemical Storage Facilities, Oct. 1978

Ontario Water Resources Act RSO 1980, c. 361

Ontario Water Resources Act: Water Management—Goals, Objectives, Policies and Implementation Procedures of the Ministry of the Environment, Nov. 1978, revised May 1984

Objectives for the Control of Industrial Waste Discharge in Ontario

Gasoline Handling Act: RSO 1980, c. 185

Gasoline Handling Code, Reg. 439/80

Guidelines for the Decommissioning and Cleanup of Sites in Ontario, January 1989

Occupational Health and Safety Act: Regulation Respecting Asbestos on Construction Projects and in Buildings and Repair Operations 654/85

Manual for an Improved Lakefill Quality Control Program, December 1988

Parameters Listing System (Palis), March 1989

Prince Edward Island

Dangerous Goods (Transportation) Act Regulations EC 319/85
Dangerous Goods (Transportation) Act SPEI 1981, c. 9
Environmental Protection Act RSPEI 1974, c. E8.1
Environmental Protection Act: Petroleum Storage Tanks Regulations PEI Reg. EC435.86

Quebec

Environment Quality Act RSQ 1977, c. Q-2
Environment Quality Act: Hazardous Waste Regulation OC 1000-85
Environment Quality Act: Quality of the Atmosphere Regulation RRQ 1981, c. Q-2, r. 20
Highway Safety Code Transport of Dangerous Substances Regulation OC 19-86
Transport Act Regulation Respecting the Transport of Waste RRQ 1981, c. T-12, r. 16
Contaminated Sites Rehabilitation Policy

Saskatchewan

Air Pollution Act RSS 1978 c. A-17
Air Pollution Control Regulation Reg. 211/75
Clean Air Regulations RRS C-12.1, Reg. 1
Dangerous Goods Transportation Act SS 1984-84 c. D-1.2
Dangerous Goods Transportation Regulations RRS D-1.2 Reg. 1
Environmental Spill Control RRS 1981 c. D-14, Reg. 1
Environmental Management and Protection Act SS 1983-84 c. E-10.2
Hazardous Substances Regulations Sask. Reg. 983/88
PCB Waste Storage Regulations RRS 1989, c. E-10.2 Reg. 6
Water Pollution Control and Waterworks Regulations RRS, c. E-10.2 Reg. 4
Environmental Assessment Act SS 1979-80 c. E10.1
Municipal Public Works Act RSS 1978 c. M-32
Waste Management Regulations Sask. Reg. 198/72
Water Quality Objectives Feb. 1977

Yukon

Dangerous Goods Transportation Act SYT 1985, c. 13
Dangerous Goods Transportation Regulation OIC 1986/118
Public Health Ordinance: Private Sewage Disposal Systems Regulations Commissioner's Order 1974-75

How to Prevent Criminal Prosecution of Senior Management

G.H. Holliday

ENVIRONMENTAL AUDITING in the United States is a way of life for major industrial organizations. In addition, auditing has become a necessity for businesses of any size prior to acquiring a facility or property.

The term "auditing" is not liked in the U.S. because the auditee associates it with financial or income tax audits, which are typically traumatic. Commonly, environmental audits within the U.S. are called (1) compliance assessment reviews (CARS) (if within an organization) or (2) environmental site assessments (ESAS) (for preacquisition audits). Having said this, I will use "audit" and "CAR" interchangeably in this paper.

From my vantage point as a consulting environmental engineer, *I find the audit program works well and is well accepted by management and field (plant) personnel alike.* In fact, given the possibility of criminal prosecution without an audit program, management and operations personnel welcome CARS.

I divide this paper into three parts: (1) the pros and cons of environmental auditing, (2) the lessons gained from environmental auditing and (3) how, in my opinion, to report an environmental audit.

Pros and Cons of Environmental Auditing

Compliance assessment reviews include positive and negative aspects. If you are being audited, you view everything as an imposition. If you are

G.H. Holliday is President of Holliday Environmental Services Inc., Tomball, Tennessee.

the auditor, you feel you are performing the functions mandated by management. Audits raise concerns regarding the findings being discovered by agencies or environmental organizations. Such discovery has not occurred in the U.S. to my knowledge. However, self reporting under air and/or water statutes has resulted in attempted prosecution. Nevertheless, in my opinion, the advantages clearly outweigh the disadvantages.

Positive Aspects of Environmental Audits

The U.S. and Canadian environmental statutes and implementing regulations allow courts to impose civil and criminal penalties on violators. Neglect by the management to establish an environmental compliance system can result in levelling criminal charges against the senior managers individually. A CAR program provides a system by which management assesses the environmental compliance management system's success or failure. Also, it provides a direct evaluation of the environmental and company policy communications system in place within the company. Finally, the audit program is a "training tool" to inform company personnel of the environmental laws, regulations, company policies and interpretations of permit provisions. Of these advantages, training is most important, since the other advantages issue from training.

I find that operating employees want to perform as management mandates. However, there must be clear and unequivocal communication and commitment from senior management to subordinates before compliance can be undertaken.

The audit program provides a continuing reminder of management's commitment to environmental compliance and the need to comply. Typically, the reminder occurs every three years with a scheduled review; however, if serious problems are observed as the result of a CAR, another audit may be requested by management within six to twelve months.

Negative Aspects of Environment Audits

Once management informs the staff of the company intention to comply with environmental requirements and establishes an adequate CAR program, senior management basically is exempt from criminal prosecution resulting from neglect of duty. This, in effect, places the compliance responsibility at a lower level, i.e., superintendents and foremen. (Once these employees understand they can go to jail, however, they become avid supporters of the audit program.)

A concern often expressed regarding compliance assessment reviews is legal exposure associated with a bad environmental audit report. I agree, the management is vulnerable, if the reported deficiencies are ignored and/or the violation is not corrected. I am willing to go before any judicial body, however, and defend an organization which has in place an environmental management system, an environmental audit program and a follow-up system to correct the deficiency even when deficiencies are found. Such a system evidences "due diligence," protecting management from prosecution.

Auditing Experience Gained within the U.S.

Environmental auditing commenced in the U.S. about ten years ago. Initially none of us knew how to approach an audit of this type. Senior management had mandated the program but had not supported it vigorously. Middle management saw the auditing program as a "black hole" into which to pour money without increasing income. Foremen and superintendents were aware that their immediate supervisors were not supporting the program, and simply tolerated the intrusion of an audit team.

Additionally, management used the environmental group as a place to "pocket" engineers who could no longer advance into management positions. In other words, initially environmental engineers were "second class corporate citizens." Today, management vigorously supports environmental programs, requests audit results and ties salaries to environmental and safety compliance. Additionally, environmental counsel is sought from company environmental engineers.

A number of lessons have been learned during the intervening years of auditing:

Lesson 1: Senior Management Must Support the Audit Program

Gradually, senior management realized they could and would be held criminally responsible for environmental deficiencies caused by the lack of an environmental management system and a CAR program. CAR programs today are fully supported by senior management. As a result, middle management supports the system. Superintendents and foremen, finding their salary progression depends on environmental compliance and the audit findings, fully support the audit program and use it to their advantage to train their employees. Also, management uses environmental engineering as a valuable training position for "upwardly mobile" personnel because of the valuable and wide-ranging training the position provides. A company is better off without a CAR program, however, if the program is only a "paper tiger" without management support.

Lesson 2: Use the Environmental Audit Program to Train Employees

Some organizations consider foremen incapable of understanding environmental regulations. In fact, some organizations currently isolate their first line supervisors from compliance knowledge. This, in my opinion, is a mistake which can cause environmental non-compliances and resultant bad press.

Environmental regulations have a direct impact on field or plant operations. Accordingly, the action of a foreman, purposely deprived of environmental know-how, can inadvertently result in civil penalties being imposed on the management or supervisor. Also, lack of environmental knowledge can result in criminal charges levelled against the first-line supervisor.

Thus, the environmental audit program should have as its primary function environmental compliance training. This is not to say review of records and facilities is not important, particularly since the audit provides senior management with the means of establishing whether the environmental information generated within the company reaches the first-line supervisor. Training occurs during the entrance and exit interviews.

Typically, an audit includes:

1. Entrance review—training session
2. Record review
3. Facility review
4. Exit interview—review of findings and additional training
5. Formal report to management.

The training is particularly useful during the exit interview when the supervisor can relate the audit findings to the environmental regulations discussed during the entrance interview.

Lesson 3: Divide Audit Findings into Various Classes according to the Seriousness

CAR findings should apply to federal, state and provincial, and local regulations and *company policies.* These findings can be serious, resulting in actual or potential civil or criminal violations. Typically, such audit findings are called "deficiencies" and in environmentally aware organizations, deficiencies are reported to the company president who requires immediate explanation and correction.

Some findings represent "concerns" since the problem can lead to substance releases or non-compliances, if not corrected. Typically, concerns are included in the formal report and must be explained or corrected by the audit organization.

Finally, "observations" are typically reserved for incomplete non-mandated records, housekeeping problems or lack of adequate signing. Normally, observations are not included in the formal report.

Remember, the audit team is there to help—not to say "I gotcha." Accordingly, discuss the shortcomings and explain the reason they will be included within the report.

Lesson 4: Don't Leave without Informing Management of the Findings and Recommendations

The exit interview should be used to inform the local management of the audit findings. I leave a handwritten review of the applicable regulation or company policy and a description of the deviation observed. Also, I leave a recommendation for correction of the finding.

This step serves several purposes. First, it provides an opportunity for foremen to correct any misobservations on the part of the auditor. Believe me, you will make mistakes in the heat of the battle. Second, it assures the foreman that there will be no surprises when the formal report comes to him for response. Third, the handwritten report provides a record to remind the foreman of what corrections will be mandated by the formal report.

Lesson 5: Establish a Formal Follow-up System

The audit findings must be acted upon or refuted by the responsible supervisor. This typically is accomplished by a formal report which passes from the corporate environmental manager to the responsible first-line supervisor, that is, to the field foreman or engineering manager, via the chain of command. The response goes back up the management chain until the manager of environment receives assurance that the findings have been refuted or corrected.

Deficiencies, i.e., serious problems, and ultimately their resolutions are sent to the chief operating officer for his information in a summary report, typically, on a quarterly basis. However, many company presidents require immediate reporting of non-compliant findings.

The above procedures assure prompt and complete corrective action and normally generate a procedure to avoid reoccurrence. Don't allow the findings to go uncorrected by neglect. Such lack of action encourages more non-compliance.

Lesson 6: Don't Forget to Train the Design Engineer

Invariably, air permit non-compliances result from installing a compressor or sweetening unit or modifying a gas plant without recognizing the new equipment or process revision triggers a permit requirement. In every case, Engineering did not know a permit was required: Engineering is left out of the loop. Consider installing a new project inventory system which is reviewed at the weekly staff meetings attended by the environmental manager. The appearance of a new project proposal on the inventory alerts the environmental manager of the need for review of the project for permit requirements. Also, the environmental manager should sign-off on the project to assure it complies with environmental regulations. It is much easier to initiate a permit application prior to commencement of construction than after the facility has operated while when you must go "hat in hand" to request a permit after the fact.

How to Report Audit Findings

Two types of reporting methods are available:

1. a "privileged document" which is sent by the senior auditor to an attorney. The attorney is either "in-house" or retained counsel. This establishes a client attorney privilege which *may or may not protect the document from discovery.* The attorneys like this approach

because of the financial income without the need to directly participate in the audit, and
2. a formal document prepared by the senior auditor and sent to the environmental manager responsible for the audit program.

The non-privileged document allows for wide distribution and review of the findings. The privileged document typically is restricted to six copies and very limited distribution. Accordingly, I favour the non-privileged document, as do most of my clients.

In my opinion, any finding which can be revealed by "discovery" can be found by agency field observation or substance release information. Accordingly, nothing is gained by attempting to conceal the formal report unless, the audit program is only a paper tiger which attempts to gain the impression of assuring compliance. Then you are better protected without the CAR program.

Some companies have expressed concern over the agencies using the report against the company to extract an administrative penalty. Experience indicates that if agencies take such an approach, companies abandon the auditing program, with a resultant increase in non-compliance. Accordingly, I have not witnessed "report seizure" unless the deficiency was located by an onsite audit by an agency.

Summary

An environmental audit program is a valuable management tool for companies serious about environmental compliance. Conversely, the audit program increases criminal liability for the management of companies not interested in complying with the environmental laws. In either event, public opinion will support the compliant company and public opinion and financial penalties will destroy the non-compliant company.

Environmental Auditing at Shell Canada

T.E. Wood

EVERY SPEAKING COURSE I have ever been on tells me never to get into my topic right away. The audience will spend the first two minutes of your presentation evaluating you without listening to you anyway, so why not wait for a bit and get into your topic when their minds have tuned back in. Fortunately for me, this process of having you evaluate me falls right into the topic I want to speak to. Right now, you are reviewing *your* standards for speakers—they should be of such and such appearance, they should have clear voices, loud enough to be heard—a multitude of factors. As soon as you have established these criteria you apply them to me.

You are auditing me, and so, since we're both into the same task I can launch into a topic near and dear to my heart—environmental auditing, and specifically, what we at Shell have learned from our corporate auditing program. I'll grant you that there is a tremendous potential for this to be one of the monumentally dull speeches in this entire symposium. After all, I doubt that anyone here wants to know the specific locations at any of our refineries where the bituminous covers on the dyke walls have broken down due to foot traffic. On the other hand I could describe my exploits as a sort of environmental Sherlock Holmes, snooping around Shell facilities, rubbing my hands gleefully as I find smoking vents and stinking wastewater streams, and anticipating dragging down staff to atone for their sins. I don't intend to do either of these things. What I do intend to do is review our lessons in developing an audit mentality, how to organize and conduct the audit itself, and how to live with a completed audit.

T.E. Wood is Staff Engineer, Shell Canada Limited, Calgary.

Whenever I talk to any of our site-staff members, I emphasize that auditing is more like a visit to a dentist or brake shop, and not a witch hunt. Most times the news is good, a few minor repairs are needed but the basic structure is sound. And even when the news is bad—your brakes are likely to fail in the next block—the consequences of knowing and acting on the information to correct the problem far outweigh the cost of the repairs. In addition to this, there is an increasing burden of due diligence placed on staff at all levels that requires us to *seek out* problems and correct them. It is not acceptable to hide your head in the sand.

Proper auditing is an opportunity to define standards for the company, reinforce existing good activities, and an opening to upgrade substandard conditions. After an audit the site knows where it stands, and can formulate a stages program to correct problems in a cost-efficient manner. The auditing staff develop a broad understanding of the problems and solutions facing the sites and can pass the information along. All really good results. Before I go on too much further extolling the beauties and benefits of audits, I'd like to toss out a warning. *There is a downside.* Audits can create embarrassing and painful situations if they are not handled properly.

Let's look at it from this negative point of view for a while. What an audit does is gather together, in one neat package, all of the problems that exist at the site. It goes further and identifies what should have been done on the basis of good industrial practice, company policy, and legislated requirements. Put yourself in the position of an environmental officer charged with finding out problems, correcting them, and taking companies to court. Wouldn't *you* like to find that kind of summary, all that work done, and with the company's own staff backing it?

So what should a responsible company do? On the one hand, audits are necessary to ensure that you're not developing really ugly environmental situations. They're required to prove that you are duly diligent in trying to prevent problems. On the other hand, they provide the best opportunity ever to shoot yourself in the foot.

The first thing you do is determine why you want to do the audit and what information you want to gather. Next, you must determine what you will do with the data and reports once they're available. This is also the stage to start to figure out what your stance will be if something really unpleasant turns up. Lastly, the distribution of data and confidentiality of the audit must be identified and agreed to. If the staff you expect to do the audits has not had previous experience in auditing or formal training, it might be wise to get them brought up-to-speed before letting them out in the field. Give some thought to the type of people you have doing the audit. A cool human calculator of an engineer might be able to tell you all about the thickness of the sewer pipes at the water treatment plant, but maybe you'd prefer to have someone that can talk with the operators and find out that they don't think that management supports them in trying to run the environmental treatment systems. Maybe you'll conclude that the audits should be conducted by teams featuring both of these types, plus old process operators that have seen every method of covering up problems, and lab people that understand quality control and so on.

Rather than discuss audits in the abstract, let's go through how I conduct audits for Shell Canada.

Step 1. The manager of the site to be audited requests that I conduct an audit for him. This establishes his ownership of the audit. The request is followed up with a conversation reviewing the Shell Environmental Audit Program, emphasizing the aspects of confidentiality and the need to develop an action plan after the audit to address identified problems.

Step 2. Acting on this request I prepare a pre-visit questionnaire which will provide me with a good base of information on the environmental status of the site. These questions cover the topics of an environmental management, site operation, and physical facilities. In cases where I am completely unfamiliar with the operation, I will visit on a informal basis to see what goes on, the size, location and staffing. This accomplishes two things. First, I get a general flavour of the area, and secondly, the questions I ask in the questionnaire don't make me look like a fool. I usually have an audit team in mind by this time, drawing on staff from head office and field locations. I personally like using field staff. If their site has problems they sometimes find solutions at the audited location. If the audited site has problems that their home site has solved, the information gets passed on.

Step 3. The pre-audit questionnaire is distributed to the site with instructions that the completed questionnaire be sent back uncopied with all input material gathered to answer the questions returned to its original location.

Step 4. The audit team and I review the questionnaire, identifying areas that need further investigation. We work up a site schedule indicating either the people or the positions we want to talk to. Both the schedule and list are flexible as, once you are on the site, your perceptions and needs can vary.

Step 5. The team and I visit the site for about a week. We review environmental controls at the site as perceived by management, as interpreted by the employees, and as they actually exist. We interview staff at all levels, tour the facilities, and poke and pry. Generally, before we leave we are in a position to review the findings of the audit with the site management team.

You might have noticed that we do not do sampling or testing as part of our audits. In this regard, the audit is more of a systems review than a physical review. As an example, the audit will not identify problems with groundwater, but it will identify the shortcoming—"The site has no way to detect problems with groundwater."

Step 6. We write up the finding as a formal report. The report details, in tabular format, the items (good and bad) that were identified together with an indication of the standards used to evaluate the situation, recommended actions to be taken, and a prioritization of the need to address the problems. The report is reviewed by site management for factual errors and by legal staff for unfortunate turns of phrase.

In producing the audit report it is necessary that the audit staff have any access to notes taken in the field and reference material obtained through the pre-questionnaire. After the request is written, all auditors' notes are discarded as they are generally filled with *non sequiturs,* notes to the auditor himself and other trivia. This material is generally useless when taken out of context (or within a few weeks of the audit when the auditor has moved on to other work).

Step 7. The report is issued. The site gets the original report. Only Legal and Environmental Affairs have copies of the report.

Step 8. The site must then prepare an action plan to address the problems identified as requiring action. In some cases the action plan may be a statement that I, as auditor, did not understand what I was seeing; everything is really fine and no action is required. In other cases,the plan may be to initiate studies and to follow up these further studies with activities to address the problem.

This step is critical to the whole audit process. Remember how I said that there is nothing an environmental investigator would like to find more than a neat tidy report detailing all of the problems at a site referenced to the appropriate company policy, permit criteria, or legislation. I lied. There is something he would like more: to find that package, to know that the site management was aware of all those problems and knew how to address them, and then find out that they did nothing at all to correct the problem.

Step 9. The site is re-audited periodically to ensure that the action plan is being followed and that no new problems are developing.

In conclusion, I would like to restate my two principal thoughts on environmental auditing and the lessons that we've learned. First: environmental audits are a positive and practical tool to reinforce a company's environmental policies. Properly conducted and acted on, the program will underscore well-run programs and provide an opportunity to upgrade problem areas in a proactive and cost-efficient manner. And second: the potential for audit-related liabilities during investigations by enforcement staff is real but it can be managed. If proper precautions are taken then the advantages aspects will far exceed this problem.

Summary of Ottawa Proceedings

K. Shikaze

CONGRATULATIONS! You have sat through two days of intensive effort focused on environmental audits. This, in a time when the preoccupation, particularly by the media and the public at large, is with other environmental issues such as toxics, climate change, ozone, and wastes. It is encouraging to see that level of interest in environmental auditing which I consider to be a very proactive means to resolve some of the issues of today. What I would like to do is to review the past few days, and I propose to do this by asking a number of questions: Who are we? What did we hear? What did we not hear? What did we achieve? What are we going to do with what we have heard? Finally, I would like to carry out a brief environmental audit of the symposium itself.

Who Are We? Based on a review of the registration, I notice that we have representation from government agencies, industries both large and small including sectors dealing with chemicals, petroleum, food, mining, steel, power generation, communication, and transportation. Also among us are consultants, academics, educators, students, lawyers, native organizations, and if I have left anybody out, my apologies. I am particularly heartened by the level of representation from government agencies of which there are a large number and, by my count, over twelve departments and agencies are represented. If there is any shortcoming it has been a lack of broader representation from the smaller industries.

What Did We Hear? We heard a number of case histories and examples of environmental auditing, and I will not report or give you details of

K. Shikaze is Regional Director General, Environmental Protection, Environment Canada, Toronto.

these, since it would be rather presumptuous on my part. We heard that environmental audits can be broad and narrow, can address regulatory compliance or can deal with company policies. We heard about links to sustainable development, concerns about confidentiality; we heard that it was a management tool and a good tool; that it was an investment and cost dependent. We also heard that it was not a panacea and there are some down sides in so far as problems are identified and there will be costs associated with rectifying those problems. In the final analysis, environmental audits can be said to encompass many functions and have many characteristics and that they are unique to the facility for which they are being undertaken.

What Did We Not Hear? All of the subjects related to an environmental audit were covered. Some were covered in more detail than others; however, there was some repetition and for those of us who have heard this subject discussed for the first time, I believe this is useful. For others who have sat in on similar sessions, it may have been redundant, but it is always useful to be reminded. Some of the specifics that I'd like to address, and which we did not hear, relate to quality control and the question of who audits the auditors. Another issue we did not hear about in any great detail was how much an audit costs, how do we plan for it and budget for it, whether the costs should be internalized or whether there should be a special budget allocation. Another area that was mentioned but not treated in any detail related to education and training. I believe there are many benefits associated with environmental audits; perhaps these subjects could be the focus of some future meetings. Another factor related to education is the need to promote environmental audits further, not only within the organization, but externally, and the question of how this might be improved could be more fully addressed.

What Did We Achieve? One of the objectives of this symposium was to promote environmental audits in government and industry. By the broad representation that we had at the symposium, I believe we have advanced the state-of-the-art of environmental audits. We have heard from the U.S. experts and have had the benefit of hearing about experiences in the U.S. and Canada showing what has worked and what has not. We have heard that environmental auditing is not a panacea and that it is a dynamic process.

What Are We Going To Do With What We Have Heard? I trust that most of you will not pay lip service to the term "environmental audit" in the future. If you do, then your time at this symposium has been wasted. In some cases, because most of you are here because you are interested, not because you are forced to come, we may be preaching to the converted. Suffice to point out to you, however, that you are now experts in the environmental audit area, and I trust that you are now committed to the process. You are a very select group who hopefully will be able to use environmental audits and will be able to act on the results. You are also missionaries (if I may call you that) so that the current level of conducting environmental audits of 25 percent will be significantly increased in the future. This you can do by applying peer pressure, not only within your organization, but externally. By doing so, you will become part of the solution, not part of the problem.

Those are the comments I have with respect to the symposium itself. I indicated to you that I would carry out a short environmental audit of the symposium itself. I would like to focus it on environmental factors and health-related matters.

On the positive side, after a delayed start, the coffee breaks were very good and positive since they not only provided coffee, both caffinated and decaffinated, but tea and juices as well as muffins, to rejuvenate ourselves during the course of the day. Vegetarian meals were provided to those who wished them and there was no smoking during the course of the symposium. In terms of policy aspects, French translation provisions were provided. Conference organizers were careful to ensure that china cups were used rather than disposables; there were provisions for the handicapped, and in general the room conditions were comfortable and conducive to an attentive audience. The presentations themselves were excellent and the presenters are to be congratulated.

On the negative side, there are some things I would like to point out. Recycled paper was not used; there was no provision for office paper recycling; the proceedings in the handouts were printed on only one side of the paper, and finally, the timing was bad because the symposium dates coincided with the visit of former U.S.S.R. president Mikhail Gorbachev, which significantly disrupted traffic in the Ottawa area.

Finally, on a positive note, the host-bar that was held proved to be a very useful mixer and enabled all of us to meet and talk to each other and exchange views on environmental audits in a relaxed atmosphere.

I would like to congratulate the organizers of the conference and those who supported it, namely The Banff Centre, Ontario, Quebec, Alberta, and the federal government. I am pleased that this symposium was held in the Ontario region and pleased to be a part of it.

Summary of Calgary Proceedings

J. Swiss

BEFORE STARTING MY summary remarks, I would like to comment on some of the points Clem Pelletier made yesterday regarding public awareness of and interest in environmental issues. Over the last two months, two experiences have demonstrated this to me in a very direct way.

About two months ago, I attended a lecture that was given by Dr. Kenneth Hare, a global-warming guru from the University of Toronto. It was held at the Jubilee Auditorium and I expected perhaps one hundred or maybe two hundred people at most would attend. But when I got there it was obvious that someone had booked a rock concert at the same time because the parking lot was full and there were people everywhere. I was wrong. All of those people were there to hear Kenneth Hare. The Jubilee was full. I thought that was amazing.

The second experience occurred yesterday when I came to this symposium. When originally asked to make this presentation, I somehow got the impression that the meeting was being held in Banff. The Banff Centre must use this as a technique for getting people to attend. So I had pictured myself delivering this closing address to maybe 20 or 30 people sitting around a table in Banff and when I got there and saw the number of people in attendance I was, again, amazed.

I think that both of these little experiences are graphic examples of some of the statistical information Clem presented yesterday showing the increased awareness and interest in environmental matters. This interest and awareness, I believe, is driving the whole process.

J. Swiss is Senior Coordinator, Health, Safety and Environment, Canadian Petroleum Association, Calgary.

I would like to thank the organizers of this symposium for asking me to give this closing address. I feel like a bit of an imposter, however, because although I have worked in the environmental field for many years and have done some small Transportation of Dangerous Goods audits, I have never actually been involved in the type of large-scale environmental audit discussed here over the last two days. This may, at first glance, seem to put me at a disadvantage in doing this wrap up, but in some respects it provides me with the advantage of observing and commenting on the environmental auditing process from a totally objective point of view. Besides, it seems that these days absolutely everyone is an environmental expert, so that should give some license to provide you with my comments today.

But where to start? I did not want to provide a repetition of everything you have heard over the last two days. So the approach I selected was as follows: First, I will review the theme and objectives of the symposium. Then I will provide a summary of what I feel were some of the major points made by speakers yesterday and this morning. I will not summarize the findings of the interactive workshops, because the rapporteurs have done a good job at this and the information should be fresh in your minds. I will comment on how well the symposium addressed its theme and objectives. Lastly, I will provide some summary thoughts. To start with, I will recapitulate the theme and objectives of the symposium.

THEME

The Environmental Challenge of the 1990s
Upgrading Environmental Auditing Practices in Canada

Objective of the Symposium

1. The symposium will provide a forum of international exchange of policies, practices and trends on environmental auditing by:

2. Advancing the state-of-the-art of environmental auditing in Canada.

Very laudable goals! Let's see how we did!

First of all, some major points that were made by speakers yesterday and this morning, with some of my own personal observations added in for good measure:

There were a number of definitions of environmental audits given. They all had similar elements, but were not exactly the same. The one I liked best was as follows:

> The environmental audit identifies and assesses past, present and potential environmental problems, verifies compliance with regulations, and corporate standards and assures that corporate action is implemented.

I think that the definition contained all of the elements that would result in a good environmental audit.

It was noted by a number of speakers that environmental auditing should be looked at as an opportunity to define standards for the company, reinforce existing good practices and act as an opening to upgrade substantial conditions. But, to succeed, the process has to have the full support of senior management and include all other employee levels as partners in the process.

There seemed to be general agreement that industry associations such as CPA, CCPA, CPPA and PACE have a role to play in environmental auditing. This role is related primarily to the development of guidelines and standards for environmental auditing.

Another point—environmental audits should:

- be company initiated
- be systematic using preplanning and forethought
- measure performance against established standards
- provide for verification of results.

Several major trends seemed to be developing in environmental audits, including:

- "top down" audits to ensure present and continued compliance
- the development of/linkage between environmental audits and risk management and insurance functions
- expansion of environmental audits beyond fencelines (e.g., contractors, suppliers and waste handlers)
- and increased use of environmental audits during pre-purchase assessments.

There seems to be increasing debate regarding environmental auditing as a public policy tool versus its use as a private management tool. There is a trend to institutionalize auditing by making the whole process more standardized and formal.

There also seems to be considerable concern regarding the legal ramifications of doing environmental audits. Although speakers from both the Canadian provincial and federal governments and U.S. governments promoted environmental auditing as an internal management tool, there was also an implicit, and in some cases explicit, indication that audit results could and would be used in court.

A number of speakers described the mechanics of doing environmental audits, including pre-audit activities, field work and providing audit results. There seemed to be considerable concern that speakers were repeating themselves, covering old ground. But I thought that it was interesting to see the similarity of approaches taken by speakers from such widely varied backgrounds. I think that this indicates that the environmental audit process is gelling, maturing and developing standards of its own.

Several speakers mentioned the importance of reviewing audit results with site management prior to leaving the site. Some provide information through verbal debriefings, whereas others provide a written review of what will be the major findings and recommendations. In either case, the importance of such feedback from a morale and partnership point of view cannot be overemphasized.

The case histories or special applications presentations moved the discussion from the theoretical plain to the "real world." I think that it is always instructive to show how environmental management tools such as the environmental audit process actually work when given the acid test of application to actual worksites. Hearing these case histories, the successes, the pitfalls, the lessons learned, was very instructive.

In summary, I think that this has been a very good symposium. The agenda was well thought-out; the speakers were excellent; and a lot of thought-provoking information on environmental audits was presented. I feel that the symposium has met its objectives, but I would like to leave you with a few additional thoughts.

At present, the environmental audit process seems to be flexible and adaptable to a wide variety of industry and government needs. I feel that this flexibility is good and should be monitored. There is a danger in allowing the process to become too rigid and formal.

Second point: There seems to be a basic contradiction built into the process. On the one hand, there is a desire to make environmental audits, as one speaker put it, "user friendly." Yet, at the same time, there is an implicit threat that audit results will reflect negatively on company operations or possibly even be used in court. Governments should resist the temptation to move towards the legalistic use of these valuable management techniques and should instead encourage their development as more effective environmental protection tools.

Finally, I think that we are at a crossroads or point of decision regarding the development of the environmental audit process. I feel that we have an opportunity to design a process that will truly suit industry and government needs. Let's not make the mistake of sliding, by default, into a legalistic environmental audit process used solely for the purpose of complying with legislation. Let's develop this powerful technique to allow us to manage our industry activities by taking full account of their environmental implications.

APPENDICES

Appendix I: Group Work

Workshop 1: Audit Pitfalls and Potentials

WHAT HAVE BEEN some of the common problem areas confronting you and how have you overcome them?

Topics to be Covered

How should one set performance standards? Who sets the standards in the company? How do standards of auditing differ from standards of performance?

Management Expectations

How should one deal with different levels of management who have different expectations? What are the needs of the audit? How should one set up the communications for the audit? How does facility management view the audit? What training skills are needed? How is the audit team selected?

Calgary Group Work

Pitfalls

Most companies have policies, but written performance standards are needed. Standards include regulations, cultural norms and understood practices. Who sets the standards: a cross-section of the company, or field management? One must also understand standards of auditing (versus performance standards). It is crucial to communicate process, standards, objectives, and outcomes.

Management Expectations

Be clear on objectives. Different levels of management (senior, middle, facility) have different expectations. Consider needs: are they for compliance, improvement, credibility with the public, accountability (personal and corporate), due diligence? Communicate what the audit can and cannot do, the audit's role as a management tool, and the cost of compliance, and communicate these points to all levels.

Facility Management

Facility management may feel threatened and imposed upon. In addition, audits create work and demand for (unavailable) resources. Sell the positives, and be sensitive to field needs. Senior management should sell the audit process. It is useful to keep in mind the carrot/stick approach.

Training

Audit skills must be cultivated. Training (basic and refresher) is critical. Consider who is benefited by certification.

Team Selection

Selection of a team depends on objectives and scope. Needed are auditing skills, operation experience, and people skills. Independence is important. What is involved will differ within large versus small companies. Personal conflicts must be resolved. The team should involve a cross-section from the company. Another factor is continuity on the part of a corporation's environmental department and pool of auditors (these groups cross-fertilize each other).

Ottawa Group Work

Pitfalls

One pitfall is failure to clearly define the *purpose* of the audit, leading to inability to set the scope. General purposes for *ongoing* operations identified over the session are: compliance, management efficiency and efficacy, and risk reduction over the long term. Special, more focused purposes have been identified as: real estate transactions, consumer interests, and marketing strategy. A further pitfall is failure to communicate the identified purpose of the audit to middle management and line-level people. A third is failure to establish and identify to all participants the standards to be used in the audit. Clarity in standards is particularly important concerning emissions or effluents, management practices, and standards for risk assessment. A fourth pitfall is failure to adhere to the original stated purpose. This may result from questionable ethics, a change of client management, or a change of government.

Workshop 2: Ensuring the Audit Program Meets Your Objectives

QUALITY IS ASSURED through the process you go through to set up audit programs. It should be a management tool. What are company audit standards? How do we establish objectives and scope of the audit? How do we measure audit quality? What are the necessary auditor credentials/ qualifications? How to deal with quality control of the audit? Post-audit evaluation of benefits.

Calgary Group Work

Audit Standards

Auditors must establish objectives and scope of the audit (i.e., compliance and management systems). They should clearly establish the measurement criteria or standards to which they decide to audit (these include protocol establishment).

Measuring Audit Quality

Set up the methodology (i.e., protocols, documentation, team composition).

Auditor Credentials/Qualifications

The scope of the audit dictates the required credentials of audit team members. Specific attributes include knowledge of process, environmental issues and management process. A team leader is necessary.

Quality Control—Audit the Audit

At issue here are consistency, quality control, effectiveness, impartial assessment, evolution of the process and allowing for the training component. Two levels are involved: report level versus process level.

Post-Audit Evaluation of Benefits

Management involvement is needed to close the loop, to avoid the audit report's becoming a shelved document (i.e., action plan). How do you judge the usefulness of the audit, and what are its benefits? What is needed here is confirmation of action plan implementation. Audits can identify good aspects of operations as well as bad, and there is opportunity to build on good aspects within the company. If an audit isn't acted upon, liability becomes a potential issue.

Ottawa Group Work

Mandate and Terms of Reference of Audit Group

The audit group needs management support (for instance, in obtaining adequate resources). Concerning the expectations of management, of whatever level, consider: (1) problems only or extent of problems, (2) compliance only/*risk* assessment, and (3) acceptable industry practice. As well, know who the audience is.

Audit Objectives, Scope and Criteria

Objectives are set by senior management, policy objectives. Scope is limited: inclusions and exclusions are clear, involving specific and precise wording. The following are standards against which the audit would measure: compliance, policies, public issues, and code.

Audit Standards and Protocols

One standard is the company code of practice (which is related to the company environment). Another is environmental policy and compliance requirements. One should also consider GAAA's.

Credentials and Qualifications

Audits require the joint expertise of facility experts, audit field, environmental specialists and generalists, insiders, and corporate persons. Other attributes of importance to the audit team are credibility, independence, the project leader, no taint of conflict of interest, training, and diplomacy.

Quality Control

Quality control is of great importance. Concerning mechanisms: self-assessment is okay; external assessment is *desirable;* the affected party should be involved; management input/feedback is needed. Benchmarks are: follow-up, resources, independence, timing, scope, recommendation priorities, standards, and objectives. Also needed are a scoring system, consistency, a positive audit, and some means of ensuring that the audit program meets your objectives.

Workshop 3: Field Techniques and Aids to On-Site Audits

HOW DO YOU, as an auditor, collect information? How can this be made more efficient? How to schedule the audit? What should the checklists look like? Other on-site data collection issues are considered.

Calgary Group Work

Scheduling

Schedule in advance in conjunction with the site. Make the schedule rigid at first, but once on the site be flexible. Remember the time frames and the importance of efficiently collecting information.

Checklists

Be specific. Checklists benefit auditors, increasing their efficiency.

Protocols

How should one conduct an audit? What works and what doesn't? Consider: company directives (who, when, how, what), report distribution, and external agents (government).

Record Review

Make use of "random" as well as prior review; ensure that records are in one place and accessible; do not allow photographs; pay attention to turnover/morale, untrained personnel, continuity. Technical sampling is not normally included.

Entrance Meeting

Fine-tune the schedule; ensure that all involved parties are present at the meeting; put the schedule in writing.

Interviewing

Interviewing requires personal skills and listening skills (both verbal and non-verbal), trust, comfort, a one-on-one or group supervisor (junior versus senior), experience in plant operations and in the audited workplace, and confirmation.

Daily Documentation

Daily documentation should be organized and formal. Summarize it daily (ask: is immediate action required?) Access to such documentation should be restricted, and photographs or videos with all copies should go to the "client" upon receipt of the final report.

Record Keeping

Maintain personal notes, co-working papers. Consider who gets reports, as well as the number in circulation.

Other

Training—at all levels—should reduce turnover; efficiency and effectiveness of environmental audits.

Ottawa Group Work

Group A

The group considered field techniques and aids to on-site analysis, with the purpose of creative, directed conversation. The product was a ten-minute presentation of the group's views.

Questions asked were: How do you as an auditor collect information? How can this be made more efficient? How to schedule the audit? and What should the checklist look like?

Protocols

What works and what doesn't? How to conduct an audit. The following actions have worked in the past: (1) Use a ten-member team whose members are knowledgeable in the field, and (2) tour the site together and split into smaller discussion groups; the group did this, and spent three days in random informal questioning.

Before Audit

Develop the scope of the audit (i.e., details, activities of the work program). Look at the applicable regulations, corporate policies, and site-specific guidelines. Do a pre-audit visit and/or submit a questionnaire or review: site maps and diagrams, organization/reporting structure, process/flow, certificates of approval.

During Audit

Confirm key site personnel for interviews.

Efficiency of the Audit

How can you make an audit more efficient?

1. List information to examine, i.e., lab records, engineering reports, operating logs, MIS.
2. Keep good field notes.
3. Prepare questions ahead of interviews.
4. Write protocols: used detailed protocols as a guide.
5. Audit team review findings daily.
6. Use an experienced person to assist team.

Protocol

What works, and what doesn't?

1. One protocol for each major area (e.g., air emissions, safety).
2. Protocols should be detailed: reference legislation, internal and corporate policies, good management practices.
3. Use protocols to ensure all areas are covered in the audit.
4. For a generic protocol, review to determine applicability to site.

Scheduling an Audit

1. Give adequate notice and schedule around availability of key personnel at site (short notice; small firm—longer notice; larger firm).
2. When you are scheduling for a site visit, schedule interviews for each day (avoid conflicts in individual's time). Examples are Kirchaff Management and Tours/interviews/preliminary assessment/exit meeting.

What Should the Checklist Look Like?

1. Collecting information: Information needs to be determined by audit type and scope. Split information needs: pre-audit on-site. Pre-audit needs are site maps, diagrams and historical records; organizational and reporting structure; and a review of legislative regulations and internal policy. On-site needs are a tour, interviews, and a review of records.

2. Improving efficiency: Advance questionnaires; protocols as guides; pre-schedule and prepare for interviews; limit/match the time period under review to audit objectives; limit record review for time and efficiency.
3. Scheduling the audit: Give adequate notice and consider availability of key personnel; block out set time for the audit.
4. On-site: Provide a detailed schedule (as soon as possible); co-ordinate interviews; employ daily team meetings to refine schedule.
5. Protocols: The relative merits of the following need to be considered: case-by-case vs. generic protocols; internal vs. consultant; and detailed vs. simple checklists.
6. Other on-site data collection issues: pictures and interview techniques.

Group B

How do you as an environmental auditor collect information efficiently? What works? What doesn't work?

Areas Considered

1. Pre-audit activities: what information should be provided to the site—objectives (senior management and site management) and scope (conveyed to individual site); information to obtain from site; flow sheet of process; pre-site visit; staff commitment and assistance; base line date; policy and regulations; define on-site resources, coordinator.
2. Developing the audit team—methodology and checklist: Alternatives are to set checklist open-ended combination; interviews with management and staff; tours; examine training; areas to examine are air, water—effluents, ground water, waste, and spills (response).
4. Additional information needs: Records of training, incidents, non-compliance, maintenance (equipment and calibration), monitoring of data, permits, and waste.
5. Management systems/protocols: Policy and standards; roles and responsibilities; training, job descriptions; preventive maintenance; contingency plans (ER); incident investigation and reporting.

Workshop 4: Government/Industry/Public Roles in Auditing

WHAT ARE THE roles of each of the sectors in the auditing field? What is the role of the third parties? What should the focus of an environmental audit be? Is cooperation between government and industry required? If so, what are the areas of cooperation? Where do we go from here? Liability of audits? Liability of consultants?

Calgary Group Work

General Observations

Can poor industry involvement in a government/industry workshop be attributed to lack of industry interest in cooperation? More likely the answer is that industry recognizes the value in environmental audits. Evidence is apparent in a number of industry presentations. Industry focus is on pragmatic "get on with business" attitude in other workshops. It is inadvisable to confine the focus of environmental audits to compliance alone (e.g., waste minimization or enhancement opportunities). The group also considered compliance on the part of smaller industries which don't have industry associations, and raised concerns regarding information access versus confidentiality.

Is Cooperation Required?

The initial reaction would appear to be an obvious yes! Cooperation is required on the part of more than just government and industry. Public interest in obtaining access to the process and to information has increased dramatically. However, cooperation is not as easy as it sounds. Each group has its own biases! Depending on the situation, public access may be a given (e.g., municipal government audits versus industry). Government is interested in audits which are largely industry driven but is satisfied with the fact that they are being done.

Areas of Cooperation

Both government and industry can encourage, sponsor workshops and assist in training. Government can develop generic guidelines (e.g., Environment Canada, EPA, etc.). Industry associations (e.g., CPA) are already playing a leading role in providing direction. Government generic guidelines are of particular value to smaller industries without formal associations.

Where Do We Go?

Environmental auditing is still in its infancy—it is too early for regulations. Industry associations should therefore continue to play a leading role. Continued examination of the issue of audit distribution versus confidentiality is called for. The feeling was that government should access industry documents by legal means. Continued cooperation in undertaking workshops which provide for interchange with government, industry and public was recommended.

Ottawa Group Work

Group A

Government's role is to set clear standards (global, local), both environmental and in auditing; *ensure compliance* with these standards; promote leadership in environmental protection. Facilitators should balance carrot and stick/incentives and penalties; passive not mandatory/right to audit necessary; consistent enforcement surveillance (level playing field); research needed into setting standards, what is achievable (technology), precise impact (science, not perception).

Industry's role is to protect interest in ensuring continued existence, compliance to regulation standards; protect the environment; and protect the health of employees and residents (value of property—not liability). Manuals exist to assist management, and management should follow them; employee environmental awareness and liability; savings from audit; report to shareholders; public goodwill—environmental performance report. Industry associations set standards and policies. Should they be self-regulating? Commitment involves regularity, an acceptable standard, commitment resources (correct, improve), and willingness to address the public's concerns.

The public's role is to press for change (for environmental improvement protection); convey concerns to government (regulators) and individuals (operators, government); seek reassurance (requires feedback). Labour unions (not environmentally safe); environmental groups. Confidential reports? Environmental assessments and environmental auditing are not adversarial: they are cooperative and have the same purpose—protection of the environment and health.

Group B

Group B considered government/industry roles in environmental auditing. The actors are: government (federal, provincial, municipal/local), industry, associations, individual companies, third parties (consultants, the public [general, specific, interest]. The audit team consists of consultants, officials, and government inspectors.

Potential Roles

The potential role of government is regulatory (standards). It involves policy, guidelines, funding (helping small companies with incentives/disincentives), licensing and quality control (using existing inspection mechanisms, e.g., health and safety). Industry's role concerns internal standards and meeting government requirements. The role of third parties is that of potential actors (e.g., Junction Triangle—community role independent auditors).

Collaboration

Use existing industry/government mechanisms; set baseline procedure with industry associations; feedback on effectiveness (QAQC); education.

The Role of Government

Government's role is to show leadership through self-auditing (as with industry, there must be a buy-in from the top to the bottom of the organization).

Where Do We Go From Here?

Canada needs to go beyond U.S. audits of big companies. Small companies need to be brought onside (e.g., companies could draw on industry/agriculture, associations for assistance; for example, tax breaks).

Workshop 5: Managing Legal Issues

WHAT ARE THE ways to manage the potential legal risks and consequences? The risks—doing an audit versus not doing an audit. The scope of the audit—what is an audit? Privilege of confidentiality, senior management commitment, public information. Issues are: How to conduct an audit; how to scope the audit; solicitor-client privilege; senior management commitment; public information.

Calgary Group Work

Conducting an Audit

Are there legal risks? There is no defence for due diligence; material accessible without control—confidentiality; failure to identify liability; identify liability not previously known.

Scoping the Audit

Consider jurisdictions, specific sites, applicable legislation, internal procedures and practices—checklist.

Solicitor-Client Privilege

Potentially more honest answers; basis of protection from exposure, if later needed.

Senior Management Commitment

This matter relates to due diligence.

Public Information

Public access to information can be considered as a risk management issue; it has an employee component.

Ottawa Group Work

Managing Legal Issues

The group agreed that there are significant benefits to auditing, which serves a range of purposes, from the policy level to the operations level (which helps to define what issues come into play, and resolutions)

Major Issues

Major issues are: double jeopardy (self-incrimination), what constitutes "due diligence"), the responsibility of auditors, and the imposition on existing management systems. Management commitment is a necessity.

Solutions

It is preferable to explain benefits by using case studies, discussing potential cost savings, demonstrating good practices to public. A better position in the legal system is gained by demonstrating due diligence. Avoid buying and paying for someone else's problems. Audits are a management tool for best operation (money and time). They define liabilities (hotspots). *Action* is the key element in demonstrating due diligence. Agencies can better communicate enforcement policies. Build on existing systems.

Double Jeopardy

The gravity of the problem of double jeopardy or self-incrimination may vary across jurisdictions. The problem could be more one of perception than of reality. Benefits far outweigh this: we should not be seen as a block to auditing. Understanding increases; apprehensions decrease. The issue is an evolving one!

Workshop 6: Auditor Training Needs and Resources

WHAT KIND OF team is needed for an audit and what skills do they need? Issues are: team selection, team members, team leader, consultant's role, auditor training courses, and audit simulation training.

Calgary Group Work

Auditor Training Needs and Resources

The culture is a corporate one. The relevant needs and resources apply to any type of audit. Purpose and scope determine the skills required.

Team Selection

Characteristics of those selecting the team are: they know the audit objectives; they know the company personnel; they understand quality assurance; and they are experienced in audits. The three C's to keep in mind are consistency, continuity, and corporate compliance.

Team Members

Qualifications of team members are: regulatory knowledge, environmental science, technical and operations experience, possibly management systems knowledge, and a legal background as a resource. The team should have three–five persons, one of whom is the team leader.

Team Leader

The team leader should be a facilitator and a good organizer, have good interpersonal skills, be knowledgeable of environmental issues, and possess report writing skills. The leader could be a consultant, but a company should develop internal team leaders as soon as possible.

Consultant's Role

Key considerations are auditing experience of the company, size of the company, and independence—type of audit and public credibility. A consultant's qualifications are: professional status; ability as a trainer, facilitator, and team leader; ability to provide technical support; and possession of good report writing skills.

Auditor Training Courses

General requirements for audit training courses are: corporate policy and standards, environmental regulations, knowledge of plant processes, information flow/communication, training in audit design (objectives, steps, protocols, reporting), encouragement of industrial associations to develop training standards/programs, and linking training to existing programs (OH & S, quality assurance).

Audit Simulation Training

Pilot the audit. Use simulation (video, role playing), field walk-throughs, and protocol reviews. Train teams as teams.

Ottawa Group Work

Group A

What kind of team? Consider: terms of reference and opportunities for enhancement. The team could be an in-house audit to meet regulatory requirements. Emergency response situations may arise. Selling of audit services must be considered.

The group considered the matters of environmental impact/corporate citizenship; knowledge of operations/process; credibility/objectivity; manageable size; appropriate expertise: managerial, technical; familiarity with regulations; independence; and ensuring of economy, efficiency and effectiveness.

Team Selection/Members/Skills: multi-disciplinary; core audit group; in-house resources/consultant/balance; optimize use of in-house expertise, understanding of company mandate, generalist skills/specialist skills, interpersonal skills, professional auditor, system/policy analysts, management skills, facilitating skills, report writing, computer expertise, knowledge of regulations, quantification skills of environmental engineering. Tailor size of the audit to meet requirements. Be aware of developmental opportunities.

Team Leader: Management skills: delegate/synthesize, plan, organize and direct; credibility/leadership recognition, visibility; objectivity; knowledge of regulations; familiarity with auditing process; technical expertise; knowledge of process.

Auditor Training: is provided by Environment Canada (publications, training manuals, simulation exercises); U.S. Environmental Protection Agency; Ontario

Ministry of the Environment. Previous experience: regional municipalities, cities, other government departments, provinces. Industry associations: Canadian Chemical Producers, Canadian Petroleum Association, PACE, Canadian Standards Association. Codes: Canadian Electric Code, Environment Canada, National Master Specs. Training consultants; universities; community colleges; negotiating skills.

Appendix II: Course Schedules

27–28 FEBRUARY 1990
CALGARY CONVENTION CENTRE

Symposium Chair:
Art Stelzig
Acting Chief
Industrial Programs Branch
Environment Canada

TUESDAY, 27 FEBRUARY 1990

Morning Session: (Glenview Room)	Environmental Auditing—Where Are We Now and Where Are We Going?
	Chairman: Brian Hammond Senior Research Manager Alberta Environment
8:00	Registration (Glen Corridor)
9:00	Official Opening and Welcome Al Schulz Director Pollution Control Division Alberta Environment
9:15	Joint Keynote Address B. Burns Regional Director General Western and Northern Region Environment Canada J. Butler President Nova Corporation of Alberta

10:00	Refreshment break
10:15	Environmental Auditing as a Management Process P.T. Budzik President Peter T. Budzik and Associates Incorporated
11:00	Environmental Auditing Perspectives and Trends J.W. Reed Coordinator Environment and Safety McColl Frontenac Inc. Peter Blackall Acting Manager Environmental Quality Branch Conservation and Protection Environment Canada
12:00	Luncheon (MacLeod D) U.S. Experiences with Environmental Auditing Jim Edward Deputy Program Director Office of Federal Activities U.S. Environmental Protection Agency
Afternoon Session: (Glenview Room)	Auditing Techniques and Applications Chairman: Mark Lesky Senior Specialist Environmental Affairs Nova Corporation of Alberta
14:00	1) Pre-Audit (Planning, staffing, training, protocols) K. Kavanagh Head of Environmental Engineering Directorate Conservation and Environment Department of National Defence Ken Tsang Superintendent Environmental Services Dow Chemical Canada Inc.

14:45	2) Field Work (Data collection, defining standards) D. Chollak Corporate Environmental Advisor Canadian Occidental Petroleum Ltd. C. Pelletier President ResCan Environmental Services Ltd.
15:30	Refreshments
15:45	3) Post-Audit (Reporting, follow-up recommendations) H.P. Cleghorn Principal Consultant Concord Scientific Corporation G.L. Boucher Manager, Environmental Compliance Domtar Inc.
16:30	4) Environmental Auditing—Special Applications *i*) Our Corporate Experience with Environmental Auditing G.E. Younghusband Senior Advisor AEC Oil and Gas *ii*) Junction Triangle Audit B. Singh Environmental Relations Consultant Inc. *iii*) Municipal Treatment Plants M.G. Thorne, Acting Deputy Commissioner Works Department Municipality of Metropolitan Toronto
17:30	Reception with host bar (Garden Terrace)

WEDNESDAY, 28 FEBRUARY 1990

Morning Session:	What We Have Learned from our Corporate Audit Program Chairman: Bruce Thompson Senior Environmental Advisor Environmental Protection Environment Canada
8:00	Environmental Auditing at Shell T.E. Wood Staff Engineer Shell Canada Limited U.S. Industry Perspective and Experience with Environmental Auditing G.H. Holliday President Holliday Environmental Services Inc.
9:00	Refreshment break
9:15	Interactive Workshops 1) Audit Pitfalls and Potentials 2) Audit Quality Control 3) Field Techniques and Aids to On-Site Audits 4) Government/Industry Roles in Environmental Auditing 5) Managing Legal Issues 6) Auditor Training Needs and Resources
10:45	Refreshments
12:00	Luncheon (Garden Terrace)
Afternoon Session:	Workshops Chairman: Frank Markson General Manager Canadian Chemical Producers' Assoc.
13:00	Plenary Session—Reports from Workshop Rapporteurs
14:00	Refreshments

15:00	Closing Address J. Swiss Senior Coordinator Health, Safety and Environment Canadian Petroleum Association
15:15	Chairman's Address Art Stelzig
15:20	Symposium Concludes

29–30 MAY 1990
OTTAWA CONGRESS CENTRE

Symposium Chair:
John Shrives
Senior Program Engineer
Chemical Industries Division
Industrial Programs Branch
Environment Canada

TUESDAY, 29 MAY 1990

Morning Session: (Salon F/G)	Environmental Auditing—Where Are We Now and Where Are We Going?
	Chairman: Brian Ward Director, Southeast Region Ontario Ministry of the Environment
8:00	Registration (West Lobby/Congress Hall Level)
9:00	Official Opening and Welcome (Salon F/G) G. Cornwall Director, Federal Programs Environment Canada
9:15	Joint Keynote Address W.N. Kissick President and CEO Union Carbide Canada Ltd. R.A. Findlay Director, Ontario Round Table on Environment and Economy Guy Demers Directeur Bureau de Coordination de la Stategie Industrielle Québec Ministère de l'Environnement

10:00	Refreshment break
10:15	Environmental Auditing as a Management Process P.T. Budzik President Peter T. Budzik and Associates Incorporated
11:00	Environmental Auditing Perspectives and Trends J.W. Reed Coordinator Environment and Safety McColl Frontenac Inc. Vic Shantora Director Industrial Programs Branch Environment Canada
12:00	Luncheon (Salon B/C) U.S. Experiences with Environmental Auditing Jim Edward Deputy Program Director Office of Federal Activities U.S. Environmental Protection Agency
Afternoon Session: (Salon F/G)	Auditing Techniques and Applications Chairman: Guy Demers Québec Ministère de l'Environnement
14:00	1) Pre-Audit (Planning, staffing, training, protocols) K. Kavanagh Head of Environmental Engineering Directorate Conservation and Environment Department of National Defence Don Hames Director Environmental Sciences Dow Chemical Canada Inc.

14:45	2) Field Work (Data collection, defining standards) T.J. Fletcher Manager Safety, Health, Environment Laboratory BF Goodrich Canada Inc.
15:05	3) Post-Audit (Reporting, follow-up recommendations) H.P. Cleghorn Principal Consultant Concord Scientific Corporation
15:20	Refreshments
15:35	B. Peel Senior Engineer Alberta Power Ltd.
16:20	4) Environmental Auditing—Special Applications *i*) How Small Business Has Adopted Environmental Auditing J.M. Shepherd Vice President Nacan Products Ltd. *ii*) Junction Triangle Audit B. Singh Environmental Relations Consultant Inc. *iii*) Municipal Treatment Plants M.G. Thorne Acting Deputy Commissioner Works Department Municipality of Metropolitan Toronto
17:30	Reception with host bar (Congress Level) (Garden Terrace)

WEDNESDAY, 30 MAY 1990

Morning Session: (Salon F/G)	What We Have Learned from our Corporate Audit Program Chairman: G. Cornwall Environment Canada
8:00	Environmental Auditing at Shell T.E. Wood Staff Engineer Shell Canada Limited U.S. Industry Perspective and Experience with Environmental Auditing G.H. Holliday President Holliday Environmental Services Inc. Real Estate/Pre-acquisition Audits R.B. German Vice President MacLarentech Inc.
9:15	Refreshment break
9:30	Interactive Workshops 1) Audit Pitfalls and Potentials 2) Audit Quality Control 3) Field Techniques and Aids to On-Site Audits 4) Government/Industry Roles in Environmental Auditing 5) Managing Legal Issues 6) Auditor Training Needs and Resources
12:00	Luncheon (Salon B/C)
Afternoon Sessions:	Workshops Chairman: Jim Taylor Manager, Environmental Affairs Du Pont Canada Inc.
13:00	Plenary Session—Reports from Workshop Rapporteurs

14:30	Refreshment break
15:00	Closing Address K. Shikaze Regional Director General Environment Canada
15:15	Chairman's Address John Shrives Environment Canada
15:20	Symposium Concludes